Perspectives in geometry and topology

Ramanujan Mathematical Society Lectures Notes Series

Vol. 1: Number theory

Vol. 2: The Riemann zeta function and related themes

Vol. 3: Formal language aspects of natural computing

Vol. 4: Commutative algebra and combinatorics

Vol. 5: Convexity in discrete structures

Vol. 6: Number theory and discrete geometry

Vol. 7: Discrete mathematics

Vol. 8: Lectures on operator theory

Vol. 9: Essays on geometric group theory

Vol. 10: Techmüller theory and moduli problems

Vol. 11: Perspectives in geometry and topology

Vol. 12: The Iwasawa theory of totally real fields

Vol. 13: Advances in discrete mathematics and applications

Ramanujan Mathematical Society

Lecture Notes Series

Volume 11

Perspectives in geometry and topology

Proceedings of the Golden Jubilee International Workshop/Conference in Geometry and Topology, held at the Indian Institute of Technology, Bombay, August 2007

Volume editors

P. Sankaran
A. R. Shastri
P. Zvengrowski

International Press
www.intlpress.com

Ramanujan Mathematical Society
Lecture Notes Series, Volume 11
Perspectives in geometry and topology

Volume editors:
P. Sankaran
A. R. Shastri
P. Zvengrowski

Published in 2011 by International Press, Somerville, Massachusetts, U.S.A. under license from the Ramanujan Mathematical Society.

This work is also published in the Republic of India, exclusively by the Ramanujan Mathematical Society.

ISBN: 978-1-57146-218-3

Printed in the United States of America.

15 14 13 12 11 1 2 3 4 5 6 7 8 9

Members of the Advisory Board

Preface

This Workshop/Conference took place at the Indian Institute of Technology, Bombay, from August 17–25, 2007, in conjunction with the Golden Jubilee of this Institute. As planned, it consisted of a Workshop from August 17–21, followed by a Conference with a broad theme 'Geometry and Topology', from August 23–25, with a free day on August 22. The Scientific Committee consisted of Professors R. Kulkarni (USA and India), P. Sankaran (India), A. R. Shastri (Principal Organizer, India), and P. Zvengrowski (Canada). One of the main goals was to convene a meeting that would be of value to graduate students and younger mathematicians, as well as providing a forum for experienced mathematicians and experts in the field.

The meeting was eminently successful in achieving these goals. The Workshop portion consisted of four minicourses, presented by Professors K. Y. Lam (Univ. Br. Columbia), K. Varadarajan (Univ. Calgary), P. Wong (Bates College, Maine), and P. Zvengrowski (Univ. of Calgary). The format consisted of four 90 minute lectures daily, one for each minicourse, on each of the five days. Each lecturer had ample time to present his material in sufficient detail. Overall the material in these four minicourses encompassed a broad spectrum of important and deep parts of classical and modern algebraic topology, along with related areas in other parts of mathematics such as algebra, geometry, number theory, analysis, and applied mathematics (e.g. the theory of elasticity). It was also remarkable how many interconnections appeared between the four minicourses, as they unfolded. The specific titles of the four minicourses are Topological Methods in the Theory of Bilinear Forms (K. Y. Lam), The Wall Finiteness Obstruction (K. Varadarajan), Combinatorial and Geometric Group Theoretic Methods in Fixed Point Theory (P. Wong), and An Introduction to ℓ_2-Homology (P. Zvengrowski). Each author has put sufficient effort to make the report up-to-date.

The Conference portion consisted of about seven lectures for each of the three days (total 21), ranging from 45–55 minutes each. The lecturers included some of the most distinguished people in the field. Topics ranged over a broad spectrum, from theoretical physics to topology, geometry, algebra, etc., and included reports about some of the most recent important developments in the field such as the recent solution of the Poincaré Conjecture by Pereleman *et al.*, to name just one example.

This Proceedings consists of complete notes from the four lecture series of the workshop and reports representing a sample of the lectures given in the conference part. In addition to the subjects already mentioned, they also represent the interplay of modern algebraic topology with algebraic geometry, algebraic groups, differential geometry, and non-commutative geometry. We feel that the volume will have a lasting value for students and researchers in the specific areas that are covered. Thanks are not only due to the authors, but also to all those who generously helped in the editorial and refereeing work for the various submissions.

One somewhat unique feature of the meeting, that was greatly enjoyed by all, was the almost daily concert of Indian Classical Music, following each session. On the free day between the Workshop and Conference parts, there was a boat trip and visit to the Elefanta Caves and Temples, as well as to downtown Mumbai for a little sight-seeing and shopping.

The Organizers

Foreword

I am happy to place, in the hands of the mathematical community, this volume of the Proceedings of the International Workshop/Conference on Geometry and Topology organised on the eve of the Golden Jubilee celebrations of our institute. We hope that this volume will be of good use to students and experts in this area.

On behalf of the organizers of the event, my sincere thanks are due to all the participants and experts who have contributed to this volume. Thanks are due to my colleagues and the staff at Department of Mathematics, IIT Bombay for the support during the event. Thanks are due to the Ramanujam Mathematical Society (RMS) for coming forward to publish this volume and thanks to the technical staff at RMS for doing an excellent job of editing.

A. R. Shastri
Department of Mathematics
I.I.T. Bombay

List of Registered Participants

17th August to 25th August 2007

K. Y. Lam, Canada
Peter Wong, USA
Peter Zvengrowski, Canada
K. Varadarajan, Canada
S. M. Srivastava, ISI Kolkata
Goutam Mukherjee, ISI Kolkata
Amiya Mukherjee, ISI Kolkata
P. Sankaran, IMSc Chennai
Himadri Mukherjee, NEHU
Basudeb Datta, IISc Bangalore
C. S. Aravinda, TIFR Bangalore
M. S. Narasimhan, TIFR Bangalore
S. Ramanan, CMI Chennai
R. V. Gurjar, TIFR Mumbai
Jaya Iyer, IMSc Chennai
Swagatha Sarkar, IMSc Chennai
Shilpa Gondhali, TIFR Mumbai
Sam Johnson, Chennai
Soumen Sarkar, ISI Kolkata
Abhijit Pal, ISI Kolkata
Debashish Sen, ISI Kolkata
Mahendra Singh, HRI
Sankar Raj, IITM
S. Balaji, Chennai
Biswajit Ranasingh, NIT Rourkela
Niraj Prasad, Mumbai Uni
Ajay Thakur, IMSc Chennai
Anisa Mohamad, IMSc Chennai
S. P. Tripati, K. M. College Delhi
M. S. Raghunathan, TIFR
Nitin Nitsure, TIFR
Hemangi Shah, TIFR
B. Subash, IITB
K. Gangopadhyay, IITB
Vikram Aital, IITB
R. S. Kulkarni, IITB
Akhil Ranjan, IITB
S. R. Ghorpade, IITB
A. R. Shastri, IITB

Contents

Perspectives in Geometry and Topology No. 11, 2010, pp. 1–10.

Ricci Flow and Negative Curvature

C. S. Aravinda

TIFR Centre for Applicable Mathematics, P.O. Box 6503, Chikkabommasandra, Bangalore 560 065, India
e-mail: aravinda@math.tifrbng.res.in

Abstract. This article is an overview discussing the application of the Ricci flow technique in the situation of negative sectional curvature.

1. Introduction

In recent years, the Ricci flow introduced by R. S. Hamilton in [14] has made it possible to achieve tremendous advances in Geometry and Topology. The most remarkable of these, essentially due to G. Perelman, is a complete description of closed 3-manifolds in conformity with the geometrization conjecture of Thurston. Subsequent to Perelman's work, most of the major results that have been proven happen to be in the situation of positive curvature. The purpose of this article is to review the role played by the Ricci flow technique in the situation of negative sectional curvature.

Given a smooth Riemannian manifold M, the Ricci flow is the geometric evolution equation on the space $\mathcal{M}$ of Riemannian metrics on M. Starting with a given initial Riemannian metric g_0 on M, the goal of the Ricci flow evolution is to possibly flow towards a Riemannian metric on M that is more natural to its underlying topology; for example, if M is diffeomorphic to the 2-sphere S^2, the most natural Riemannian metric on it is the one which has constant curvature $+1$.

In higher dimensions, the so called Einstein metrics, that is, metrics with constant Ricci curvatures, are considered to be the most natural ones among all metrics; in particular, metrics of constant sectional curvature are the simplest examples of Einstein metrics. A Riemannian metric g on M is said to be an *Einstein* if $\mathrm{Ric}_g = \lambda g$, where Ric_g is the symmetric $(0, 2)$ tensor, called the Ricci tensor, associated to the metric g and λ is a constant real number; in fact, if g is an Einstein metric then the constant λ is the Ricci curvature of the metric g.

The Ricci flow (RF) introduced by Hamilton in 1982 in his seminal paper [14] is the 1-parameter family of Riemannian metrics $g(t)$ on a manifold M that satisfies the equation

$$\frac{\partial}{\partial t} g(t) = -2\,\mathrm{Ric}_{g(t)}.$$

In the same paper, Hamilton showed that given a Riemannian metric g on a closed manifold M there exists an $\varepsilon > 0$, depending on g, such that the Ricci flow equation has a unique solution $g(t)$ for t in the interval $[0, \varepsilon)$, with $g(0) = g$. At this point, two natural questions arise:

(1) Does there exist a solution $g(t)$ to the Ricci flow for all $t \in [0, +\infty)$? And, if so,
(2) Is there a Riemannian metric g_∞ on M such that $g(t) \to g_\infty$ as $t \to \infty$?

Let us consider a metric g_0 on M for which $\mathrm{Ric}_{g_0} = \lambda g_0$ for some $\lambda \in \mathbb{R}$. Then the solution to the Ricci flow equation with $g(0) = g_0$ is given by $g(t) = (1 - 2\lambda t)g_0$. If the initial metric g_0 is Ricci flat, that is, if $\lambda = 0$, then $g(t) = g_0$ is the solution to the Ricci flow; in other words, the Ricci flat metrics, of which zero sectional curvature metrics are the simplest examples, are fixed points of the Ricci flow. In particular, if $M = \mathbb{T}^2$, the standard 2-torus $S^1 \times S^1$ equipped with a flat metric g_0, then g_0 is a fixed point of Ricci flow.

If $M = S^2$ and g_0 is the metric of constant curvature $+1$, we have the evolution $g(t) = (1 - 2t)g_0$ and the sphere collapses to a point at time $t = 1/2$. This example illustrates the fact that, in general, volumes of the metrics evolving under the Ricci flow are not preserved. In fact, the volumes $\mathrm{Vol}(g(t))$ of the evolving metrics $g(t)$ satisfy the equation $\frac{\partial}{\partial t}\mathrm{Vol}(g(t)) = -\int_M R_{g(t)} d\mu_{g(t)}$, where $\mu_{g(t)}$ is the Riemannian measure of the metric $g(t)$ (that is, $\int_M d\mu_{g(t)} = \mathrm{Vol}(M, g(t))$) and $R_{g(t)}$ is the scalar curvature function of the evolved metric $g(t)$. Therefore, following Hamilton, when M is a compact n-manifold one can also work with the volume normalized Ricci flow (NRF) defined by the equation

$$\frac{\partial}{\partial t} g(t) = -2\mathrm{Ric}_{g(t)} + \frac{2}{n}\frac{\int_M R_{g(t)}\, d\mu_{g(t)}}{\int_M d\mu_{g(t)}} g(t).$$

It is easily checked that the Einstein metrics are precisely the fixed points of NRF, the normalized Ricci flow. This means that if the answer to questions (1) and (2) above is affirmative, then the manifold M admits an Einstein metric natural to its topology.

In this article, which is organized in four sections, we work mostly with the NRF instead of the RF. Section 2 deals with the case of closed 2-dimensional manifolds or surfaces and the work of Hamilton [15]. Section 3 is a brief description of Shi's work [19] and its application in Farrell-Jones topological rigidity theorem for open manifolds [7]. And, section 4 discusses in some detail the caveat about non-smooth convergence of the Ricci flow pointed out by the paper [11] of Farrell-Ontaneda.

2. The Ricci flow on surfaces

We begin by recalling the celebrated Gauss-Bonnet theorem which is, perhaps, one of the earliest results demonstrating a strong interaction between geometry and topology. More precisely, the so called Gauss-Bonnet formula

$$\int_M K\, d\mu = 2\pi\chi(M)$$

relates the Gaussian curvature K and the Euler characteristic $\chi(M)$ of a closed surface M embedded in $\mathbb{R}^3$.

In [15], Hamilton considers the NRF equation. If g is a Riemannian metric on a surface M, then $\mathrm{Ric}_g = K_g g$ and $R_g = 2K_g$, where K_g is the gaussian curvature function of the metric g. Hence, NRF simplifies to the equation

$$\frac{\partial}{\partial t} g(t) = (r - R_{g(t)}) g(t)$$

where $r = \int_M R_{g(t)}\, d\mu_{g(t)}$. Furthermore, the Gauss-Bonnet formula implies that $r = 4\pi \chi(M)$ and consequently, for a closed surface M, we see that r is a constant independent of the metric $g(t)$. Or, if one writes the above equation in local coordinates, it looks like

$$\frac{\partial}{\partial t} g_{ij}(x,t) = (r - R_g(x,t)) g_{ij}(x,t).$$

Observe that if $K < 0$ then the Gauss-Bonnet theorem implies that $\chi(M)$ is negative and hence M is a surface of genus bigger than 1. In this case, Hamilton proves the following theorem [15, Theorem 1.2]:

Theorem 2.1. *If $r < 0$, for any arbitrary initial data, the solution $g(t)$ to NRF exists for all positive time, is unique and converges as, $t \to \infty$, to a metric g_∞ of constant curvature.*

Remark 2.2. Theorem 2.1 is also proven for the cases $r = 0$ and $r \geq 0$, (cf. [15] and [3]), but the proofs are a bit different and involved for these cases; since our emphasis in this article is on negative curvature, we shall not discuss them here.

As a consequence of Theorem 2.1 and Remark 2.2, one obtains a new proof of the classical uniformization theorem for Riemann surfaces (see [2] for a more recent discussion). The uniformization theorem says that every smooth surface admits a metric of constant curvature. And, the above theorem says that there exists a metric of constant negative curvature, say -1, on a closed surface of genus bigger than 1. We give below a brief sketch of the proof.

Since a Riemannian metric g completely determines the various geometric invariants like the covariant derivative, the curvature tensors, the curvatures etc., the evolution of g under the NRF determines the evolution of its scalar curvature R_g. The evolution equation for $R_{g(t)}$ turns out to be the following simple equation:

$$\frac{\partial R_{g(t)}}{\partial t} = \Delta_{g(t)} R_{g(t)} + R_{g(t)}^2 - r R_{g(t)} \tag{1}$$

where $\Delta_{g(t)}$ is the Laplace operator associated to the metric $g(t)$. Recall that the Laplace operator Δ_g, for the metric g, acting on functions is defined by $\Delta_g f = \frac{1}{\sqrt{|g|}} \sum_{i,j=1}^n \frac{\partial}{\partial x^i}\left(\sqrt{|g|} g^{ij} \frac{\partial f}{\partial x^j}\right)$, where $|g| = \det(g_{ij})$ and g^{ij} is the matrix inverse of g_{ij}.

This type of equation is known as a *reaction-diffusion equation*. On the right hand side, the Laplacian term promotes diffusion of R, whereas the quadratic reaction terms promote concentration of R. If there was only the Laplacian term, then the equation would be the heat equation. If there were only the quadratic reaction terms, then the equation would be an ODE and the solution would blow up in finite time for any initial data satisfying $R(0) > \max\{r, 0\}$. The answer to the question of how the scalar curvature evolves under NRF depends on which term dominates. Under certain curvature assumptions, like positivity of Ricci curvature in dimension 3 (see [14]), it is possible to show that the diffusion term dominates, thereby allowing one to employ tools similar to the ones used in solving the heat equation.

The main tool used by Hamilton in proving Theorem 2.1 is the so called Maximum principle, one version of which that is used in the case of heat equation is the following:

Maximum principle. Let $u : M \times [0, T) \to \mathbb{R}$ be a C^2 solution to the heat equation $\frac{\partial u}{\partial t} = \Delta_g u$ on a closed Riemannian manifold. If $C_1 \leq C_2$ are two constants such that $C_1 \leq u(x, 0) \leq C_2$ for all $x \in M$, then $C_1 \leq u(x, t) \leq C_2$ for all $x \in M$ and $t \in [0, T)$.

A physical interpretation of the above maximum principle is that if one starts with an arbitrary heat distribution on a conducting material, then the maximum and minimum temperatures at a point at any evolving positive time cannot exceed those at the starting time. Using the heat equation analogy for the case of heat equation with nonlinear terms, like the equation (1), Hamilton showed that, for any solution of NRF on a closed surface M of genus bigger than 1, there exists a constant $C > 0$ depending only on the initial metric, such that $r - Ce^{rt} \leq R_{g(t)} \leq r + Ce^{rt}$ holds for all $t \in [0, T)$. Observe here that $r < 0$ since M is a surface genus bigger than 1, and this estimate on $R_{g(t)}$ implies that if the solution to NRF exists for all $t > 0$ then $R_{g(t)} \to r$, a negative constant, as $t \to \infty$.

Indeed, the same estimate can also be used to conclude the existence of long time solution for NRF as follows. If $T = \infty$, there is nothing to argue for. Otherwise, if the curvatures of the evolved metrics under NRF are bounded for $t \in [0, T)$, then it can be shown that all the higher order covariant derivatives of the Riemann curvature tensors are also bounded for $t \in [0, T)$; this means that as $t \to T$ the metrics $g(t)$ converge to a Riemannian metric $g(T)$ on M. With $g(T)$ as the initial data now and using the short time existence of the solution to NRF, one gets a solution on an interval $[T, T_1)$; uniqueness of the solution then guarantees that this solution smoothly extends the solution to the original NRF beyond time T and up to the time T_1. If $T_1 = \infty$, we are done; otherwise, repeat the same process and extend the solution beyond T_1. This way, using the above estimate, one shows the existence of the long time solution for NRF. As pointed out earlier, this proves Theorem 2.1.

3. Ricci flow on noncompact, complete manifolds

In this section, we quickly recall the work of Shi [19] on the Ricci flow (RF) deformation of metrics on noncompact manifolds and then give a very brief description of its application in the Farrell-Jones topological rigidity theorem for open manifolds [7].

A first step in the study of evolution of a flow governed by a partial differential equation is to see that there exists a unique short-time solution starting from a given initial data.

For the evolution of the Ricci flow on a closed Riemannian manifold M, Hamilton used the heat equation analogy and the powerful Nash-Moser implicit function theorem to prove the short time existence and uniqueness of the solution starting from an arbitrary initial metric. This was reproved by DeTurck in [4] who used a certain argument, now known as *DeTurck's trick*, to transform the Ricci flow equation on a closed manifold M into a strictly parabolic equation for which existence and uniqueness of the solution for short time follows from the heat equation methods. And, solving a certain ODE, one obtains the short time existence for the Ricci flow.

When the manifold M is noncompact, a lot more effort is needed to prove the short time existence of a complete solution to Ricci flow. In fact, the short time existence cannot be proved for an arbitrary noncompact, complete Riemannian manifold M and Shi, who addressed this question in [19], points out that there are noncompact complete Riemannian manifolds on which the Ricci flow evolution does not have any solution for an arbitrary small time interval; this can happen, for instance, when the curvatures are unbounded. However, when the curvatures of M are bounded, the short time existence can be proven. More precisely, Shi proved [19, Theorem 1.1] the following short time existence theorem and estimates on the higher order covariant derivatives of the Riemann curvature tensor:

Theorem 3.1. *Let $(M, g_{ij}(x))$ be an n-dimensional noncompact complete Riemannian manifold with its Riemann curvature tensor R_{ijkl} satisfying $|R_{ijkl}|^2 \leq k_0$ on M, where $0 < k_0 < \infty$ is a constant. Then there exists a constant $T(n, k_0) > 0$ depending only on n and k_0 such that the Ricci flow evolution equation has a smooth solution $g_{ij}(x, t)$ for a short time $t \in [0, T(n, k_0)]$.*

Furthermore, for any integer $m \geq 0$, there exist constants $C_m > 0$ depending only on n, m and k_0 such that

$$\sup_{x \in M} |\nabla^m R_{ijkl}(x, t)|^2 \leq C_m / t^m, \qquad 0 \leq t \leq T(n, k_0).$$

A Riemannian manifold is called *A-regular* if for some sequence $A = \{A_m\}$ of nonnegative numbers, we have

$$|\nabla^m R_{ijkl}| \leq A_m$$

where the indices i vary over the natural numbers and $\nabla^m R_{ijkl}$ is the m-th covariant derivative of the curvature tensor. Note that the 0-th condition means that the sectional curvatures are bounded, that is, bounded away from $\pm\infty$.

Theorem 3.1 of Shi shows that any complete Riemannian metric g with pinched sectional curvature can be approximated by a complete A-regular metric $\overline{g}$. Furthermore, if the sectional curvatures of g are contained in the interval $[a, b]$, then, given $\varepsilon > 0$, the metric $\overline{g}$ can be found so that its sectional curvatures are contained in $[a - \varepsilon, b + \varepsilon]$ (see also [17]). This fact was used by Farrell-Jones to prove their topological rigidity theorem for compact nonpositively curved manifolds with non-empty boundary. More precisely, Farrell-Jones proved the following theorem ([7, Theorem 0.1]):

Theorem 3.2. *Let M be a compact aspherical manifold with possibly non-empty boundary. Suppose that* $\dim M \neq 3, 4$ *and $\pi_1 M$ is isomorphic to the fundamental* *group of an A-regular complete non-positively curved Riemannian manifold N. Then M is topologically rigid.*

Recall that a closed manifold M is *topologically rigid* if any homotopy equivalence $h : N \to M$ from another manifold N is homotopic to a homeomorphism. A compact manifold M with boundary ∂M is *topologically rigid* if any homotopy equivalence $h : (N, \partial N) \to (M, \partial M)$ where N is another compact manifold and the restriction $h|_{\partial N}$ is a homeomorphism that satisfies the following:*There exists a homotopy* $h_t : (N, \partial N) \to (M, \partial M)$ *from h to a homeomorphism with* $h_t|_{\partial N} = h|_{\partial N}$. (When $\partial M = \emptyset$, this agrees with the notion of topological rigidity already defined above.)

Remark 3.3. For closed, aspherical manifolds M, with $\dim M \neq 3, 4$, topological rigidity theorem was proven by Farrell-Jones (see [5]) when N is a closed nonpositively curved manifold. The case of dimension 3 is now known to be true; that is, if $f : M \to N$ is a homotopy equivalence between closed 3-manifolds such that N supports a Riemannian metric of nonpositive curvature, then f is homotopic to a homeomorphism. D. Gabai proved this in [13] when N is hyperbolic, provided the Poincaré Conjecture is true. When N is nonpositively curved, the geometrization conjecture implies that either N supports a hyperbolic metric or N is sufficiently large in the sense of Waldhausen. And, Waldhausen proved the result in [20] for sufficiently large aspherical 3-manifolds provided the Poincaré Conjecture is true. Now that the solutions to both the classical Poincaré conjecture as well as the geometrization conjecture are widely accepted to be correct, the dimension 3 case of topological rigidity follows from the above discussion. The case of dimension 4 is still open.

It may be worth recording here that Shi's theorem has also been used in a paper of Kanai [16] who proves that cusps of finite volume negatively curved manifolds with strictly $\frac{1}{4}$-pinched curvatures are diffeomorphic to $T \times [0, \infty)$, where T is a compact flat manifold.

4. Non-smooth convergence of the Ricci flow

In this section we describe a work of Farrell-Ontaneda [11], showing the existence of pinched negatively curved metrics for which the normalized Ricci flow (NRF) does not converge smoothly.

We first recall that, in [21], R. Ye studied the NRF when the initial Riemannian metric g is negatively curved and proved that a negatively curved Einstein metric is strongly stable; that is, the NRF starting near such a Riemannian metric g converges (in the C^∞ topology) to a Riemannian metric isometric to g, up to scaling.

In the same paper, R. Ye also proved that Riemannian metrics on a closed manifold with curvatures pinched sufficiently close to -1 can be deformed, through the NRF, to hyperbolic manifolds, but the pinching constant in his theorem depends on other quantities (e.g the diameter or the volume). Ye's paper was motivated by the problem on whether the NRF can be used to deform every metric with curvatures pinched sufficiently close to -1 to an Einstein metric (the pinching constant depending only on the dimension).

Definition. *The NRF for a negatively curved Riemannian metric g is said to converge smoothly if the NRF, starting at g, is defined for all t and converges (in the C^∞ topology) to a well defined negatively curved Einstein metric.*

It is a consequence of Ye's paper [21] that the NRF for a negatively curved Riemannian metric g converges smoothly if and only if the NRF, starting at g, eventually gets into a stable neighbourhood of some negatively curved Einstein metric g_∞. A *stable neighbourhood of g_∞* is a neighbourhood for which any NRF starting there converges to a metric isometric (up to scaling) to g_∞. These stable neighbourhoods can be taken as sufficiently small open sets just in the C^2 topology. Using the notion of stable neighbourhoods, Farrell-Ontaneda proved the following theorem in [11].

Theorem 4.1. *Given $n > 10$ and $\varepsilon > 0$ there is a closed smooth n-dimensional manifold N such that*

(1) *N admits a hyperbolic metric*
(2) *N admits a Riemannian metric g with sectional curvatures in $[-1-\varepsilon, -1+\varepsilon]$ for which the NRF does not converge smoothly.*

Before giving a sketch of the proof of this theorem, let us briefly say a few words about its background. Soon after proving their topological rigidity theorem for closed negatively curved manifolds, Farrell-Jones showed in [6] that smooth rigidity fails, in general, by constructing negatively curved metrics on exotic smooth structures on certain closed real hyperbolic manifolds M^n, $n \geq 7$; these exotic smooth structures were obtained by taking connected sums of M^n with homotopy spheres Σ^n.

Later, P. Ontaneda showed in [18] that smooth rigidity fails in dimension 6 by constructing negatively curved metrics on exotic PL-structures on certain closed, 6-dimensional real hyperbolic manifolds; this construction was generalized in [8] for all dimensions bigger than 6. The exotic PL constructions are more powerful in the sense that they not only demonstrated in [9] that, in general, the unique harmonic map in the homotopy class of certain homotopy equivalences between two closed, PL-nonhomemorphic negatively curved manifolds need not be a homeomorphism, but that such harmonic maps exist (see [12]) which are not homeomorphisms even between two diffeomorphic closed negatively curved manifolds. In all of these constructions, one of the manifolds could be taken to be real hyperbolic.

Strengthening the exotic PL constructions further, Farrell-Ontaneda showed in [10] that there exists a harmonic map between closed negatively curved Riemannian manifolds M and N which is not a diffeomorphism but is the limit of a 1-parameter family of diffeomorphisms; a map which is the limit of a 1-parameter family of homeomorphisms is known as a *cellular map*. The theorem proved by Farrell-Ontaneda in [10, Theorem 1 and its Addendum] is the following:

Theorem 4.2. *For every integer $n > 10$, there is a harmonic cellular map $h : M \to N$, between a pair of closed negatively curved n-dimensional Riemannian manifolds, which is not a diffeomorphism. Indeed, the map h can be approximated by diffeomorphisms, either M or N can be chosen to be a real hyperbolic manifold and the other chosen to have its sectional curvatures pinched within ε of -1, where ε is any preassigned positive number.*

They also remark that it is not clear whether the map h constructed in Theorem 4.2 is a homeomorphism or whether a harmonic homeomorphism between closed negatively curved Riemannian manifolds must always be a diffeomorphism.

Theorem 4.2 is actually an easy consequence of the following proposition, also proven in [10].

Proposition 4.3. *Given an integer $n > 10$ and a positive number ε, there exist n-dimensional closed, orientable manifolds M_0, M_1, N, Riemannian metrics g_0, g_1 on M_0 and M_1, respectively, and smooth finite covers $p_0 : N \to M_0$, $p_1 : N \to M_1$ with the following properties:*

(1) *M_0 and M_1 are homeomorphic but not PL-homeomorphic.*
(2) *The metric g_0 on M_0 is real hyperbolic; that is, has constant sectional curvature -1.*
(3) *The metric g_1 on M_1 has sectional curvatures in the interval $[-1-\varepsilon, -1+\varepsilon]$.*
(4) *There is a C^∞ family of C^∞ metrics h_s on N, $0 \le s \le 1$, with $h_0 = p_0^* g_0$ and $h_1 = p_1^* g_1$, such that every h_s has sectional curvatures in the interval $[-1-\varepsilon, -1+\varepsilon]$.*

The manifold N of Theorem 4.1, is actually the one constructed in Proposition 4.3. We now briefly indicate a proof of Theorem 4.1, using Proposition 4.3.

First observe that the metric h_0, being the pull-back metric of g_0, is also real hyperbolic. And, since the NRF preserves isometries, we have that if the NRF on N with initial metric $h_1 = p_1^* g_1$ converges smoothly then the NRF on M_1 with initial metric g_1 also converges smoothly. Therefore, assume that the NRF starting with the metric g_1 converges smoothly (for, otherwise we are done!) and let $g_{1,t}$ be the solution to the NRF starting with $g_{1,0} = g_0$ and converging to a negatively curved Einstein metric $g_{1,\infty}$.

Next, it follows from the strong rigidity theorem of Mostow that if $g_{1,\infty}$ is real hyperbolic then it is necessarily isometric to the real hyperbolic metric g_0; but, because of property *(1)* of Proposition 4.3, this is impossible and hence $g_{1,\infty}$ is not real hyperbolic. This implies that the metric g_1 isn't real hyperbolic either and consequently, the metrics $p_1^* g_1$ and $p_1^* g_{1,\infty}$ also aren't real hyperbolic.

Now, if the NRF does not converge smoothly starting with the metric h_s for some s, we are again through. Therefore, assume that the NRF starting with metric h_s converges smoothly for each $s \in [0, 1]$. Let $h_{s,t}$ be the solution to the NRF starting with the metric $h_{s,0} = h_s$ and converging to the negatively curved Einstein metric $h_{s,\infty}$.

Since h_0 is real hyperbolic we have that $h_{0,t} = h_0$ for all $t \in [0, \infty]$. And, since every negatively curved Einstein metric is stable (see [1, p.357]), we can assume that all negatively curved Einstein metrics $h_{s,\infty}$ have neighborhoods V_s for which any NRF starting in V_s converges to a metric isometric (up to scaling) to $h_{s,\infty}$ (see [21, p.873]).

Since the NRF depends continuously on the initial conditions in the space of Riemannian metrics with the C^∞ topology, it follows that the function $(s, t) \to h_{s,t}$ is continuous for $s \in [0, 1]$, $t \in [0, \infty)$. Therefore, every $s \in [0, 1]$ has an open neighbourhood I_s such that $h_{s',\infty}$ is isometric (up to scaling) to $h_{s,\infty}$ for all $s' \in I_s$. This means that in the quotient space $\mathcal{N}/scaling$, of all Riemannian metrics on N

modulo scaling by positive constants, the map $s \to [h_{s,\infty}]$ from the interval [0, 1] to $\mathcal{N}/scaling$ is locally constant and hence is continuous (here $[h_{s,\infty}]$ denotes the equivalence class of $h_{s,\infty}$). This is a contradiction because $h_{0,\infty} = h_0$ is real hyperbolic and $h_{1,\infty}$ is not real hyperbolic and the Theorem 4.1 follows.

Acknowledgements

§4 of this article is based on a talk given at the Department of Mathematics, IIT Bombay as a part of their Golden Jubilee conference during August 2007. I wish to thank A. R. Shastri and P. Sankaran, both for the invitation to speak at the conference and for the suggestion to write it as an article. I also thank P. Sankaran for a thorough reading of the preliminary version of the article and for suggestions to make it more readable. Thanks are due to Tom Farrell for clarification regarding Remark 3.3.

References

[1] A. L. Besse, Einstein Manifolds, Ergebnisse Series vol. **10**, Springer-Verlag, Berlin (1987).

[2] X. Chen, P. Lu and G. Tian, A Note on Uniformization of Riemann Surfaces by Ricci Flow, *Proc. Amer. Math. Soc.*, **134** (2006) 3391–3393.

[3] B. Chow, The Ricci flow on the 2-sphere, *J. Differential Geom.*, **33** (1991) 325–334.

[4] D. M. DeTurck, Deforming metrics in the direction of their Ricci tensors, *J. Differential Geom.*, **18** (1983) 157–162.

[5] F. T. Farrell and L. E. Jones, Topological rigidity for compact nonpositively curved manifolds, Proc. Symp. Pure Math., vol. 54, Part 3, pp. 229–274, Providence, R.I.: Amer. Math. Soc. (1993).

[6] F. T. Farrell and L. E. Jones, Negatively curved manifolds with exotic smooth structures, *J. Amer. Math. Soc.*, **2** (1989) 899–908.

[7] F. T. Farrell and L. E. Jones, *Rigidity for Aspherical Manifolds with $\pi_1 \subset GL_m(R)^*$*, **2** (1998) 215–262.

[8] F. T. Farrell, L. E. Jones and P. Ontaneda, Hyperbolic manifolds with negatively curved exotic triangulations in dimensions larger than five, *J. Differential Geom.*, **48** (1998) 319–322.

[9] F. T. Farrell, L. E. Jones and P. Ontaneda, Examples of non-homeomorphic harmonic maps between negatively curved manifolds, *Bull. Lond. Math. Soc.*, **30** (1998) 295–296.

[10] F. T. Farrell and P. Ontaneda, Harmonic cellular maps which are not diffeomorphisms, *Invent. Math.*, **158** (2004) 497–513.

[11] F. T. Farrell and P. Ontaneda, A Caveat on the Convergence of the Ricci Flow for Pinched Negatively Curved Manifolds, *Asian J. Math.*, **9** (2005) 401–406.

[12] F. T. Farrell, P. Ontaneda and M. S. Raghunathan, Non-univalent harmonic maps homotopic to diffeomorphisms, *J. Differential Geom.*, **54** (2000) 227–253.

[13] D. Gabai, On the Geometric and Topological Rigidity of Hyperbolic 3-Manifolds, *J. Amer. Math. Soc.*, **10** (1997) 37–74.

[14] R. S. Hamilton, Three-manifolds with positive Ricci curvature, *J. Differential Geom.*, **17** (1982) 255–306.

[15] R. S. Hamilton, The Ricci flow on surfaces, *Contemp. Math.*, **71**, *Amer. Math. Soc.*, Providence, RI (1988) 237–262.

[16] M. Kanai, A pinching theorem for cusps of negatively curved manifolds with finite volume, *Proc. Amer. Math. Soc.*, **107** (1989) 777–783.

[17] V. Kapovich, Curvature bounds via Ricci smoothing, *Illinois Journal of Mathematics*, **49** (2005) 259–263.

[18] P. Ontaneda, Hyperbolic manifolds with negatively curved exotic triangulations in dimension six, *J. Differential Geom.*, **40** (1994) 7–22.

[19] W.-X. Shi, Deforming the metric on complete Riemannian manifolds, *J. Differential Geom.*, **30** (1989) 223–301.

[20] F. Waldhausen, On irreducible 3-manifolds which are sufficiently large, *Ann. of Math.*, **87** (1968) 56–88.

[21] R. Ye, Ricci Flow, Einstein metrics and Space Forms, *Trans. Amer. Math. Soc.*, **338** (1993) 871–896.

Perspectives in Geometry and Topology No. 11, 2010, pp. 11–15.

Shafarevich Conjecture for Genus-2 Fibrations

R. V. Gurjar

School of Mathematics, Tata Institute of Fundamental Research, Mumbai 400 005, India
e-mail: gurjar@math.tifr.res.in

Abstract. We will discuss Shafarevich Conjecture about holomorphic convexity for smooth projective surfaces with a genus-2 fibration all whose fibers are irreducible.

Keywords. Holomorphic convexity, genus-2 fibration, fundamental group.

Introduction

There are important papers dealing with genus-2 fibrations on a minimal smooth projective surface $X/\mathbb{C}$. We mention some of the names; E. Bombieri [3], A. Beauville [2], E. Horikawa [11], A. Ogg [16], Y. Namikawa-K. Ueno [13], U. Persson [17], M. Reid [18], G. Xiao [21] and others. Our interest is in the following question:

Shafarevich Conjecture (SC). Let X be a smooth, irreducible projective variety/$\mathbb{C}$. Then the universal cover $\tilde{X}$ of X is holomorphically convex.

Recall that a connected complex manifold U is **holomorphically convex** if for any sequence of points $\{x_n\}$ in U without a limit point there is a holomorphic function f on U which is unbounded on $\{x_n\}$.

Some history.

(1) Siegel [19] proved that if $\tilde{X}$ is a bounded open set then $\tilde{X}$ can be embedded in $\mathbb{C}^n$ as a closed complex submanifold.
His result was slightly generalized by N. Mok–B. Wong [12] and J. Carlson–R. Harvey [4].
(2) R.V. Gurjar and A.R. Shastri proved (SC) for elliptic surfaces [9]. Hence (SC) is proved for all smooth projective surfaces which are not of general type. In fact Cox-Zucker [5] and Gurjar–Shastri proved a result similar to the part (1) of our main result below.
Let K_X denote the canonical divisor of X. Recall that K_X is ample if and only if $K_X \cdot C > 0$ for every irreducible curve C in X.

Jointly with B.P. Purnaprajna we have proved:

Theorem. *Let $\pi : X \to C$ be a genus 2-fibration on a smooth minimal projective surface. Assume that K_X is ample and all the fibers of π are irreducible. Then we have:*

(1) *the image* $\pi_1(F) \to \pi_1(X)$ *is trivial, where* F *is a general fiber of* π, *if* π *is not a* C^∞ *locally trivial bundle.*
(2) *(SC) is true for* X.

Corollary 1. *Let* X *be as in the Theorem. Then Madhav Nori's conjecture is true for* X. *If* $q_X = 0$ *then* $\pi_1(X) = (1)$.

Corollary 2. *Let* X *be as in the Theorem. Then* $\pi_2(X)$ *is a free abelian group.*

Preliminaries

The proof uses some properties of ruled surfaces. For example:

Lemma 1. *Let* $f : Z \to B$ *be a* $\mathbb{P}^1$*-fibration on a normal projective surface* Z. *Then* Z *has at most rational singularities. If* C *is a reduced, irreducible component of a fiber* $f^*(b)$ *and* $p \in C$ *is such that* p *does not lie on any other irreducible component of* $f^{-1}(b)$ *then* Z *is smooth at* p.

Proof of the theorem

Our idea is to study the map given by $|2K_X|$. Recall that $P_2(X) = \dim H^0(X, 2K_X) = K_X^2 + \chi(\mathcal{O}_X)$. By results of G. Xiao [21], if $K^2 > 8$ then $|2K_X|$ gives a morphism of degree 2, $\Phi : X \to Y$. By a ramified covering trick we reduce to the case when $K^2 > 8$. Since K_X is ample, Φ is a finite morphism.

A general fiber of π is a genus 2 curve, hence hyper-elliptic. By an elementary argument using $2K_X|_F$ we see the following.

Lemma 2.

(1) *If* F *is a smooth fiber of* π *then* $\Phi(F) \approx \mathbb{P}^1$.
(2) *If* F_1, F_2 *are two general fibers of* π *then* $\Phi(F_1) \cap \Phi(F_2) = \phi$.

Now Y *has a* $\mathbb{P}^1$*-fibration* $\psi : Y \to C$ *s.t.*

$$\begin{array}{ccc} X & \xrightarrow{\Phi} & Y \\ \pi \downarrow & & \downarrow \psi \\ C & \xrightarrow[1_C]{} & C \end{array}$$

is commutative. By taking normalization we will assume that Y *is normal.*

Let $F_0 = \sum_{i=1}^r m_i G_i$ *be a singular fiber of* π. *Since* $K \cdot F_0 = 2$ *by adjunction formula and the fact that* $K \cdot G_i > 0$ *we get that* $r \leq 2$. *If* $r = 2$ *then* $m_1 = m_2 = 1$. *A similar argument shows that if* $r = 1$ *then* $m_1 = 1$. *Thus every fiber of* π *is reduced. Then every fiber of* ψ *is reduced. Thus* $F_0 = G_1$ *or* $G_1 + G_2$, *where* G_i *are reduced and irreducible.*

By assumption every fiber of π is irreducible

Then every fiber of ψ is reduced and irreducible. By Lemma 1, Y is smooth and ψ is a minimal ruled fibering, i.e. a $\mathbb{P}^1$-bundle over C.

Now for a general fiber F of π, $F \to \Phi(F)$ is a $2:1$ map.

The group $H = \mathbb{Z}/(2)$ acts on F such that $\Phi(F) = F/H$.

Since Y is smooth, H has no isolated fixed points. Thus $R := X^H$ is a disjoint union of smooth curves. R is the ramification locus in X. Further $R \cdot F = 6$ for a general fiber F.

Now $Y = \mathbb{P}(\mathcal{E})$, where $\mathcal{E}$ is a normalized locally free sheaf of rank 2 on C. Let $\wedge^2\mathcal{E} = E, e = -\deg E$. Then ψ has a cross-section C_0 with $C_0^2 = -e$. Let ℓ be a fiber of ψ. Then $\mathrm{Num}(Y) = \mathbb{Z}C_0 + \mathbb{Z}\ell$. Using well-known properties of curves on Y, we get two possible cases.

(1) R is smooth, irreducible with $R^2 > 0$. Then $\pi_1(X)\tilde{\to}\pi_1(Y)\tilde{\to}\pi_1(C)$. From this we can deduce the Theorem easily (use the exactness of $\pi_1(F) \to \pi_1(X) \to \pi_1(C) \to (1)$).

(2) R is a disjoint union of 6 cross-sections $R_1, \ldots, R_6$ such that $\Phi(R)$ meets every fiber of ψ transversally ($R_i^2 = 0 \forall i$). Then $\pi : X \to C$ is a C^∞-fiber bundle.

Let G be a reducible singular fiber of π. Then $G = G_1 + G_2$, $\mathbb{Z}/(2)$ keeps G stable. Now $(G_1+G_2)^2 = 0$. Either each G_1 is stable under $\mathbb{Z}/(2)$ or permuted under $\mathbb{Z}/(2)$. We analyze G_i^2, $G_1 \cap G_2$ and the images $\pi(G_i)$. Now Y can have an A_1 singularity at $\Phi(G_1) \cap \Phi(G_2)$. Using standard intersection theory on normal surfaces we prove that the case when G is reducible cannot occur.

Now it is easy to complete the proof of the Theorem as follows.

Suppose that $\pi_1(X) \cong \pi_1(Y)$ as in (1) above. Then using the easy observation that $\pi_1(Y) \cong \pi_1(C)$ we immediately see from the exact sequence in (1) above that the map $\pi_1(F) \to \pi_1(X)$ is trivial. Let $U \to C$ be the universal covering map. Now the universal cover of X is $U \times_C X$. Using the natural complex analytic map $U \times_C X \to U$, which has compact fibers, we deduce that any sequence of points in $U \times_C X$ without a limit point maps to a sequence of points in U without a limit point. Using Uniformization Theorem for Riemann Surfaces there is a holomorphic function f on U which is unbounded on this image sequence in U. Then the pull back of f to $U \times_C X$ gives a required function on $U \times_C X$.

Now assume that π is a C^∞-locally trivial fiber bundle. Now we first take the fiber product $U \times_C X$, where $U \to C$ is the universal cover. We have a natural holomorphic map $U \times_C X \to U$. Using Bers' Simultaneous Uniformization Theorem [1] we obtain the universal cover of $U \times_C X$ as a C^∞-fiber bundle over U. If $C \cong \mathbb{P}^1$ or C is an elliptic curve then π is a holomorphically locally trivial bundle over C. In this case either $U = C$ or $U = \mathbb{C}$ and we get $U \times \Delta$ as the universal cover of X, where Δ is the universal cover of F.

If genus $C > 1$ then U is the unit disc in $\mathbb{C}$. In this case the universal cover of $U \times_C X$ is known to be a bounded domain in $\mathbb{C}^2$, hence holomorphically convex by the result of Siegel mentioned in the Introduction.

Proof of corollary 1

We indicate the argument when $B \subset X$ is a rational curve with $B^2 > 0$. Now $\tilde{X}$ is holomorphically convex. Let $\tau : \tilde{X} \to X$ be the covering map. $\bar{B} \approx \mathbb{P}^1$, hence we have lifts $\bar{B} \to \tilde{X}$. The inverse image $\tau^{-1}(B)$ is connected since $B^2 > 0$. If $\tau^{-1}(B)$ has infinitely many irreducible components then $\tilde{X}$ cannot be holomorphically convex.

If $q_X = 0$ then π cannot be a C^∞ bundle as $C \approx \mathbb{P}^1$. By part (1) of the Theorem, $\pi_1(X) = (1)$.

Proof of corollary 2

I proved a few years ago that if (SC) is true for a surface X then $\pi_2(X)$ is free abelian.

Remarks when K_X is only nef and big

In this case a singular fiber of π can have many (-2)-curves which will contract to RDP's on Y. Hence we have to work with $\mathbb{P}^1$-fibrations on rational surfaces Y having RDP's as singular points. We are hopeful that our result is still valid even in this case.

References

[1] L. Bers, Simultaneous uniformization, *Bull. Amer. Math. Soc.*, **66** (1960) 94–97.

[2] A. Beauville, L'application canonique pour les surfaces de type général, *Invent. Math.*, **55** (1979) 121–140.

[3] E. Bombieri, Canonical models of surfaces of general type, *Publ. Math. IHES*, **42** (1973) 171–219.

[4] J. Carlson and R. Harvey, A remark on the universal cover of a Moishezon space, *Duke Math. J.*, **43** (1976) 497–500.

[5] D. A. Cox and S. Zucker, Intersection Numbers of Sections of Elliptic Surfaces, *Invent. Math.*, **53** (1979) 1–44.

[6] R. V. Gurjar, Two remarks on the topology of projective surfaces, *Math. Ann.*, **328** (2004) 701–706.

[7] R. V. Gurjar and M. Miyanishi, On the Jacobian Conjecture for **Q**-homology planes, *J. Reine Angew. Math.*, **516** (1999) 15–132.

[8] P. Griffiths, Complex-analytic properties of certain Zariski-open sets on algebraic varieties, *Ann. of Math., (2)*, **94** (1971) 21–51.

[9] R. V. Gurjar and A. R. Shastri, Covering spaces of elliptic surfaces, *Composition Math.*, **54** (1985) 95–104.

[10] R. Hartshorne, Algebraic geometry, Graduate Texts in Mathematics, Springer-Verlag, Berlin (1987).

[11] E. Horikawa, On algebraic surfaces with pencils of curves of genus 2, Complex analysis and algebraic geometry, Iwanami-Shoten, Tokyo (1977).

[12] N. Mok and Bun Wong, Characterization of bounded domains covering Zariski dense subsets of compact complex spaces, *Amer. J. Math.*, **105** (1983) 1481–1487.

[13] Y. Namikawa and K. Ueno, The complete classification of fibers in pencils of curves of genus two, *Manuscripta Math.*, **9** (1973) 143–186.

[14] M. Nori, Zariski's conjecture and related problems, *Ann. Scient. École Norm. Sup.* 4e *Serie*, **16** (1983) 305–344.
[15] T. Napier, Convexity properties of coverings of smooth projective varieties, *Math. Ann.*, **286** (1990) no. 1–3, 433–479.
[16] A. Ogg, On pencils of curves of genus two, *Topology*, **5** (1966) 355–362.
[17] U. Persson, A family of genus two fibrations, Algebraic Geometry (Proc. Summer Meeting, Univ. of Coppenhagen, 1978, 496–502), *Lecture Notes in Math.*, **732**, Springer-Verlag (1979).
[18] M. Reid, π_1 of surfaces with small K^2, Algebraic Geometry (Proc. Summer Meeting, Univ. of Coppenhagen, 1978, 534–544), *Lecture Notes in Math.*, **732**, Springer-Verlag (1979).
[19] C. L. Siegel, Analytic functions of several complex variables, Institute for Advanced Study, Princeton (1949).
[20] G. B. Shabat, The complex structure of domains that cover algebraic surfaces, *Funct. Anal. Appl.*, **11** (1977) 135–142.
[21] G. Xiao, Surfaces fibrées en courbes de genre deux, *Lecture Notes in Mathematics*, **1137**, Springer-Verlag (1985).

Perspectives in Geometry and Topology No. 11, 2010, pp. 17–27.

Orthogonal Multiplications and Minimal Submanifolds of Spheres

Kee Yuen Lam

Department of Mathematics, University of British Columbia, Vancouver, B.C., V6T 1Z2 Canada
e-mail: keeyuen.lam@gmail.com

Abstract. We first establish some basic properties of orthogonal multiplications not previously recorded in the literature. Using these, we show that orthogonal multiplications, when they satisfy certain extra conditions, can be exploited to produce nontrivial examples of minimal submanifolds of spheres.

Keywords. Sums of squares, orthogonal multiplication, geodesics, minimal submanifolds of spheres.

1. Introduction

Let Λ be a commutative and associative ring with unit and $P = \Lambda[x_1, \dots, x_r, y_1, \dots, y_s]$ be the polynomial ring over Λ in the variables $x_1, \dots, x_r$ and $y_1, \dots, y_s$. For the case $\Lambda = \mathbb{R}$ or $\mathbb{C}$, A. Hurwitz in 1898 [5] asked the question: what is the minimal n such that there is an identity

$$(x_1^2 + \cdots + x_r^2)(y_1^2 + \cdots + y_s^2) = f_1^2 + \cdots + f_n^2 \tag{1}$$

for n polynomials $f_1, \dots, f_n$ in P? The study of such identities has since been vigorously pursued (see, for example [7], [8], [4] and [15]), but the minimal n, denoted by $r *_\Lambda s$ for given values of r and s, is far from being determined.

It is not difficult to see that in (1), the polynomials $f_1, \dots, f_n$ must be bilinear in $x_1, \dots, x_r$ and $y_1, \dots, y_s$ variables. For $\Lambda = \mathbb{R}$ in particular, such an identity determines a bilinear map

$$f : \mathbb{R}^r \times \mathbb{R}^s \longrightarrow \mathbb{R}^n$$

simply by setting $f(x, y) = (f_1(x, y), \dots, f_n(x, y))$, where $x = (x_1, \dots, x_r)$, $y = (y_1, \dots, y_s)$. With respect to the standard Euclidean norm, f has the "normed" property that

$$\|f(x, y)\| = \|x\|\,\|y\|$$

for all $x \in \mathbb{R}^r$, $y \in \mathbb{R}^s$. In the literature such a normed bilinear map is frequently referred to as an "orthogonal multiplication", and is regarded as a generalisation of the multiplication of real, complex, quaternionic and octonomic numbers. See for instance [15], [10] and [17]. For some basic discussion of such maps, and for their roles in topology and geometry, one can refer to [8] and [10].

Researchers in non-linear analysis are interested in orthogonal multiplications primarily because they can lead to examples of harmonic maps between Euclidean spheres, as explained in [1], [2], [3], [16], [14], [12]. The purpose of this paper is to follow up along a similar theme. We shall show that orthogonal multiplications, when they satisfy certain extra conditions, can be used to produce examples of minimal submanifolds of spheres, i.e., submanifolds whose inclusion map into the sphere is a harmonic map. For this purpose we first need to trace many of the properties of orthogonal multiplications, as established in [8] and [10], back to their most fundamental origin. This is done in section 2. The construction of minimal submanifolds of spheres is then presented in the main theorem of section 3. In section 4 we give some examples and comments.

2. Fundamental properties of orthogonal multiplications

Let $f : \mathbb{R}^r \times \mathbb{R}^s \longrightarrow \mathbb{R}^n$ be as in the introduction. It is customary to say that the orthogonal multiplication f is of type $[r, s, n]$. There is also no harm in assuming that $r \leq s$. For $x \in \mathbb{R}^r$ and $y \in \mathbb{R}^s$, we agree, as in [10], to call $f(x, y) \in \mathbb{R}^n$ the *product* of x and y, and simply denote it by xy when no confusion is likely to arise. We shall also use $\langle\ ,\ \rangle$ to denote the standard inner product in a Euclidean space. First recall

Proposition 2.1. *For $x, x' \in \mathbb{R}^r$ and $y, y' \in \mathbb{R}^s$ one always has*

$$\langle xy, x'y'\rangle + \langle x'y, xy'\rangle = 2\langle x, x'\rangle\langle y, y'\rangle.$$

This is a straightforward consequence of the polarisation identity $4\langle a, b\rangle = \|a+b\|^2 - \|a-b\|^2$, together with the fact that f is an orthogonal multiplication.

Corollary 2.2. *An orthogonal multiplication $f : \mathbb{R}^2 \times \mathbb{R}^2 \longrightarrow \mathbb{R}^3$ of type* $[2, 2, 3]$ *is always a composite $f = j\bar{f}$:*

$$\mathbb{R}^2 \times \mathbb{R}^2 \xrightarrow{\bar{f}} K \xrightarrow{j} \mathbb{R}^3,$$

where j is the inclusion of a two dimensional subspace K into $\mathbb{R}^3$. Furthermore, the orthogonal multiplication $\bar{f}$ can be identified with the multiplication of complex numbers if K and the two copies of $\mathbb{R}^2$ are orthogonally re-coordinatised in a suitable manner.

This is proposition (2.1) of [8]. For a coordinate-free explanation, choose orthonormal bases x, x' and y, y' for the two copies of $\mathbb{R}^2$. Consider the two pairs $\{xy, x'y'\}$ and $\{xy', x'y\}$ of unit vectors in $\mathbb{R}^3$. Since f is an orthogonal multiplication so that $y \perp y'$ implies $xy \perp xy'$, etc., each vector in one pair is perpendicular to each in the other pair. Consequently these two pairs of vectors span *mutually orthogonal* subspaces in $\mathbb{R}^3$. Since ambient dimension is 3, not both subspaces can be two dimensional, so one of them, say the first, must have dimension one. This means $\langle xy, x'y'\rangle = \pm 1$. But then the proposition implies $\langle x'y, xy'\rangle = \mp 1$. Hence the unit vectors $x'y$ and xy' are collinear, and the second subspace is also one-dimensional. It follows that $xy, x'y', xy'$

and $x'y$ altogether span a two-dimensional subspace K of $\mathbb{R}^3$, and f is a composite $j\bar{f}$ as claimed.

We return now to a general orthogonal multiplication f of type $[r, s, n]$. For a fixed z in $\mathbb{R}^n$ define its "left-factor space" $\mathcal{L}F_z$ to be the following subset of $\mathbb{R}^r$:

$$\mathcal{L}F_z = \{x \in \mathbb{R}^r | x = 0 \text{ or } z = xy \text{ for some } y \in \mathbb{R}^s\}.$$

Note that by our convention the zero vector in $\mathbb{R}^r$ is always considered a left-factor of z. Analogously, the right-factor space of z is the subset

$$\mathcal{R}F_z = \{y \in \mathbb{R}^s | y = 0 \text{ or } z = xy \text{ for some } x \in \mathbb{R}^r\}$$

of $\mathbb{R}^s$. In particular, $z \in \operatorname{Im} f$ if and only if $\mathcal{L}F_z$ or $\mathcal{R}F_z$ contains more than the zero vector.

As the most remarkable property of orthogonal multiplications, we now present

Theorem 2.3. *$\mathcal{L}F_z$ is a linear subspace of $\mathbb{R}^r$ and $\mathcal{R}F_z$ is a linear subspace of $\mathbb{R}^s$.*

Proof. It suffices to take $z \neq 0$ and to treat the case of $\mathcal{L}F_z$. Let x, x' be independent vectors in $\mathcal{L}F_z$. Then there exist y, y' in $\mathbb{R}^s$ such that $xy = z = x'y'$. Restrict f to the orthogonal multiplication

$$\hat{f} : \operatorname{span}\{x, x'\} \times \operatorname{span}\{y, y'\} \longrightarrow \mathbb{R}^n.$$

The two spans on the left side are two-dimensional, while the image of $\hat{f}$ lies in a subspace of $\mathbb{R}^n$ spanned by xy', $x'y$ and $x'y' (= xy = z)$. This is at most three dimensional, and by Corollary 2.2, it is actually two-dimensional. Thus we can take $\hat{f}$ to be

$$\operatorname{span}\{x, x'\} \times \operatorname{span}\{y, y'\} \longrightarrow \operatorname{span}\{z, x'y\}, \tag{2}$$

which has type $[2, 2, 2]$ and is identifiable with the multiplication of complex numbers. Since every (nonzero) complex number is a left-factor of every other complex number, each nonzero linear combination $\alpha x + \alpha' x'$ is a left-factor of z with a (unique!) right-factor of the form $\beta y + \beta' y'$, so that $\alpha x + \alpha' x' \in \mathcal{L}F_z$ as well. This proves that $\mathcal{L}F_z$ is a linear subspace. Similarly for $\mathcal{R}F_z$. Furthermore, these two subspaces have the same dimension, k, provided that $z \neq 0$. □

Theorem 2.4. *Let $f : \mathbb{R}^r \times \mathbb{R}^s \longrightarrow \mathbb{R}^n$ be an orthogonal multiplication with $f(x, y) = z = f(x^*, y^*) \neq 0$. Then as a smooth map from $\mathbb{R}^{r+s}$ to $\mathbb{R}^n$, the Jacobian matrix of f at (x, y) and at (x^*, y^*) have the same rank q. In fact $q = r + s - k$ where $k = \dim(\mathcal{L}F_z) = \dim(\mathcal{R}F_z)$.*

Proof. Since f is bilinear, its Jacobian matrix at (x, y) is given by

$$Jf_{(x,y)}(\Delta x, \Delta y) = f(x, \Delta y) + f(\Delta x, y) = x(\Delta y) + (\Delta x)y,$$

so that $q = \dim(x\mathbb{R}^s + \mathbb{R}^r y)$, with obvious meaning for the notation.

Consider the intersection $x\mathbb{R}^s \cap \mathbb{R}^r y$. It consists of vectors w in $\mathbb{R}^n$ which can be simultaneously factored as xy' and $x'y$ for some $x' \in \mathbb{R}^r$, $y' \in \mathbb{R}^s$. This leads to a type [2, 2, 2] multiplication (2) as in the proof of Theorem 2.3, which, as explained there, entails in particular that x' is a left-factor of z. Hence $w = x'y \in (\mathcal{L}F_z)y$, and we have proved

$$x\mathbb{R}^s \cap \mathbb{R}^r y \subset (\mathcal{L}F_z)y.$$

Conversely an element in $(\mathcal{L}F_z)y$ is of the form $x'y$ where $x'y' = z$ for some $y' \in \mathbb{R}^s$. From (2) again we conclude that $x'y$ has x as left-factor, so $(\mathcal{L}F_z)y \subset x\mathbb{R}^s$. Since $(\mathcal{L}F_z)y \subset \mathbb{R}^r y$ trivially, we get the reverse inclusion $(\mathcal{L}F_z)y \subset x\mathbb{R}^s \cap \mathbb{R}^r y$ as well, so that these two linear subspaces actually coincide. Similarly we have $x\mathbb{R}^s \cap \mathbb{R}^r y = x(\mathcal{R}F_z)$ also.

The Jacobian rank in question is now

$$\begin{aligned} q &= \dim(x\mathbb{R}^s + \mathbb{R}^r y) \\ &= \dim x\mathbb{R}^s + \dim \mathbb{R}^r y - \dim(x\mathbb{R}^s \cap \mathbb{R}^r y) \\ &= r + s - \dim(\mathcal{L}F_z)y \\ &= r + s - \dim(\mathcal{L}F_z) \\ &= r + s - k, \end{aligned} \tag{3}$$

which depends only on $z = xy$ and not on the way it is factored. This finishes the proof. □

It is probably relevant to point out that Theorem 2.3 and 2.4 are of interest only for orthogonal multiplications of non-classical type. The classical types, considered by Hurwitz and Radon [5], [13], are in the form $\mathbb{R}^r \times \mathbb{R}^n \longrightarrow \mathbb{R}^n$, where necessarily $r \leq 2^a + 8b$ if the integer n is equal to 2^{a+4b}(odd), $0 \leq a \leq 3$. For these types, it is obvious that $\mathcal{L}F_z = \mathbb{R}^r$ for all $z \in \mathbb{R}^n$, and f has Jacobian rank n at all (x, y) whenever $x \neq 0 \neq y$. For non-classical types however, our two theorems are nontrivial and constitute the basis for the geometric studies carried out in [8] and [10]. One can, for example, classify points z in the image of an orthogonal multiplication f according to the "degree of freedom" with which z can be factorised, namely, according to the value of $\dim\mathcal{L}F_z$. If $\dim\mathcal{L}F_z$ is r (which we assumed in the outset to be $\leq s$), then z is the "easiest" to factorise in the sense that any $x \in \mathbb{R}^r$ is a left-factor. Such z is called a "collapse point" of f in [10]. It is then immediately clear that C_f, the set of all collapse points of f, is a *linear* subspace of $\mathbb{R}^n$, say of dimension $l \geq 0$.

On the other hand one can take

$$V = \{xy | x \in \mathbb{R}^r,\ y \in \mathbb{R}^s \text{ and } \|x\| = \|y\| = 1\}$$

and prove that V is a real algebraic subvariety of S^{n-1}. Note that $V = f(S^{r-1} \times S^{s-1})$. The regular points of V are just those points $z \in V$ for which $\dim\mathcal{L}F_z$ is minimal amongst all points of V, compare [8], § 3. In a sense these are the "hardest" elements to factorise.

The collapse points of unit length are given by

$$C = C_f \cap S^{n-1} = C_f \cap V.$$

They constitute a sphere of dimension $l - 1$ linearly embedded in S^{n-1}, where if $l = 0$ we simply understand S^{l-1} to mean the empty set.

Denoting the subvariety $C_f^{\perp} \cap V$ by Z where $C_f^{\perp}$ is the orthogonal complement of C_f in $\mathbb{R}^n$, we recall from [10], Proposition 2.3, the joint-decomposition lemma below:

Lemma 2.5. *If C is nonempty, then V is the joint $C * Z$ where the joint segments are minimising geodesic arcs in S^{n-1}.*

Here is a proof of this lemma which is much more direct than that of [10]. If $c \in C$ and $z \in Z$, then z has a factorization $z = xy$ with $\|x\| = \|y\| = 1$. As a collapse point c certainly has x as a left-factor, say $c = xy'$. But then $z \perp c \Rightarrow y \perp y'$, and the great circular arc in S^{n-1} joining z to c is

$$z \cos t + c \sin t = x(y \cos t + y' \sin t), \quad o \leq t \leq \frac{\pi}{2},$$

which obviously lies in V. This shows that $C * Z \subset V$. The reverse inclusion is also easy to establish.

3. Minimal submanifolds of spheres

The construction of minimal submanifolds of spheres via orthogonal multiplications is to a large extent suggested by the work of [8] and [10], to which the present paper can be considered as a sequel. The main result is

Theorem 3.1. *Let $f : \mathbb{R}^r \times \mathbb{R}^s \longrightarrow \mathbb{R}^n$ be an orthogonal multiplication where $1 < r \leq s < n$, and let $C_f \subset \mathbb{R}^n$ be the linear subspace of collapse points. Assume that the variety $V = f(S^{r-1} \times S^{s-1})$, contained in S^{n-1}, consists of regular points and collapse points only. Let $V = C * Z$ be the joint decomposition, where $C = C_f \cap V$ is the set of collapse points in V, and $Z = C_f^{\perp} \cap V$ is as described in Lemma 2.5. Furthermore, assume that for each $z \in Z$, the orthogonality condition*

$$z \perp (\mathcal{L}F_z)^{\perp}(\mathcal{R}F_z)^{\perp}$$

is valid. Let $l = \dim C_f$ and let k be the dimension of $\mathcal{L}F_z$ for each and every $z \in Z$. Then $C_f^{\perp} \cap S^{n-1}$ is the sphere S^{n-l-1}, and Z is minimal submanifold of this sphere with dimension $d = r + s - k - l - 1$.

Remark 3.2. The orthogonality condition $z \perp (\mathcal{L}F_z)^{\perp}(\mathcal{R}F_z)^{\perp}$ can be intuitively interpreted as follows: if $a \in \mathbb{R}^r$ is perpendicular to all left-factors of z (i.e., as "remote as possible" from being a left-factor of z) and $b \in \mathbb{R}^s$ is perpendicular to all right-factors of z, then ab is perpendicular to z, i.e., ab is "as far from equal to z as possible".

Proof. By hypotheses, the complement $V - C$ consists only of regular points of the variety V, so it is a submanifold of S^{n-1}. In fact, if

$$f_1 : S^{r-1} \times S^{s-1} \longrightarrow S^{n-1}$$

is the restriction of f to unit spheres, then, mainly because of the result on constancy of Jacobian rank as asserted in Theorem 2.4, one obtains from [8], §3, that the further restriction

$$f_1^{-1}(V - C) \longrightarrow V - C$$

is a smooth fibre bundle map with base $V - C$ and fibre diffeomorphic to S^{k-1}. From the joint decomposition $V = C * Z$, one sees that $C_f^\perp$ and $V - C$ intersect transversally in $\mathbb{R}^n$, so that $Z = C_f^\perp \cap V = C_f^\perp \cap (V - C)$ is a closed submanifold in $C_f^\perp \cap S^{n-1} = S^{n-l-1}$, with dimension $d = r + s - k - l - 1$. □

We now quote [11] as reference for basic facts on minimal submanifolds.

To show that Z is minimal in S^{n-l-1}, we shall use a sufficient criterion for a d-dimensional Riemannian submanifold $Z = Z^d$ to be minimal in an ambient Riemannian manifold $W = W^m$.

Criterion 3.3. If, through each $z \in Z^d$ there pass d mutually perpendicular geodesic segments $\gamma_1, \gamma_2, \ldots, \gamma_d$ of W^m lying entirely in Z^d, then Z^d is a minimal submanifold of W^m.

Curiously enough, this criterion does *not* imply that the inclusion map $i : Z^d \longrightarrow W^m$ is totally geodesic.

The proof for this criterion is most obvious because $\gamma_1, \gamma_2, \ldots, \gamma_d$ are geodesics of Z and W simultaneously, by hypotheses. Let $T_1, T_2, \ldots, T_d$ be the unit tangent vector fields along $\gamma_1, \gamma_2, \ldots, \gamma_d$. Then the mean curvature vector τ at z, being the tension of the inclusion map i at z, is given by the trace of the second fundamental form, which in this case equals

$$\tau_z = \Delta_{T_1} T_1 + \cdots + \Delta_{T_d} T_d = 0 + \cdots + 0 = 0,$$

where $\Delta = \Delta^W$ denotes the covariant differentiation of W. This confirms that i is a harmonic map, and minimality follows.

Remark 3.4. This simple criterion can be frequently put to good use. For example, some minimal submanifolds of [11], p. 23–24, can be established quite directly by this criterion.

Coming now to our d-dimensional submanifold Z inside $W = S^m$ with $m = n - l - 1$, it suffices to exhibit, through each $z \in Z$, a total of d mutually perpendicular great circles of S^m (or of S^{n-1}) lying entirely in Z.

At any $z \in Z$ let $\{x_1, \ldots, x_k\}$ be an orthogonal basis of $\mathcal{L}F_z$, which we extend to an orthonormal basis $\{x_1, \ldots, x_k, x_{k+1}, \ldots, x_r\}$ of $\mathbb{R}^r$. Let $y_1, \ldots, y_k \in \mathbb{R}^s$ be the right-factors corresponding to the left-factors $x_1, \ldots, x_k$, so that $x_1 y_1 = \cdots = x_k y_k = z$. Since $\|z\| = 1$, $y_1, \ldots, y_k$ are unit vectors. In fact they are mutually orthogonal too. For example, to see that $y_1 \perp y_k$ it suffices to show $x_1 y_1 \perp x_1 y_k$. But by Proposition 2.1 one has

$$\langle x_1 y_k, x_k y_1 \rangle + \langle x_1 y_1, x_k y_k \rangle = 2 \langle x_1, x_k \rangle \langle y_1, y_k \rangle = 0,$$

so that $\langle x_1 y_k, x_k y_1 \rangle = -\langle x_1 y_1, x_k y_k \rangle = -\langle z, z \rangle = -1$. This implies $x_1 y_k = -x_k y_1$ as unit vectors, so $\langle x_1 y_1, x_1 y_k \rangle = -\langle x_1 y_1, x_k y_1 \rangle = -\|y_1\|^2 \langle x_1, x_k \rangle = 0$.

We can now extend $\{y_1, \ldots, y_k\}$ to an orthonormal basis $\{y_1, \ldots, y_k, y_{k+1}, \ldots, y_s\}$ of $\mathbb{R}^s$.

Consider the two sets of vectors in $\mathbb{R}^n$ given by

$$A = \{x_{k+1} y_1, x_{k+2} y_1, \ldots, x_r y_1\} \text{ and}$$

$$B = \{x_1 y_2, \ldots, x_1 y_k, x_1 y_{k+1}, \ldots, x_1 y_s\}.$$

Obviously, each set consists of unit vectors that are mutually orthogonal. We assert in addition that each member of A is perpendicular to each member of B. This can be verified in two separate steps:

Case i. For $k+1 \leq i \leq r$ and $2 \leq j \leq k$. The argument that $x_i y_1 \perp x_1 y_j$ is by now familiar; just use Proposition 2.1 exactly as in the last few paragraphs, to deduce $\langle x_1 y_j, x_j y_1 \rangle = -1$ and hence conclude that $x_1 y_j = -x_j y_1$. It follows that $\langle x_i y_1, x_1 y_j \rangle = \langle x_i y_1, -x_j y_1 \rangle = -\|y\|^2 \langle x_i, x_j \rangle = 0$.

Case ii. $k+1 \leq i \leq r$ while $k+1 \leq j \leq s$. Note that we have chosen $x_1, \ldots, x_k$ to span $\mathcal{L}F_z$ and $y_1, \ldots, y_k$ to span $\mathcal{R}F_z$, so $x_i \in (\mathcal{L}F_z)^\perp$ and $y_j \in (\mathcal{R}F_z)^\perp$. By hypotheses, $x_i y_j \perp z (= x_1 y_1)$, or $\langle x_i y_j, x_1 y_1 \rangle = 0$. Proposition 2.1 again gives

$$\begin{aligned} \langle x_i y_1, x_1 y_j \rangle &= 2\langle x_i, x_1 \rangle \langle y_j, y_1 \rangle - \langle x_i y_j, x_1 y_1 \rangle \\ &= 0 - 0 = 0, \text{ as required.} \end{aligned}$$

Now take an orthonormal basis $c_1, \ldots, c_l$ of the linear space C_f of collapse points of f. By definition of "collapse point", they must all have x_1 as left-factor. Let us write, with slightly artificial indexing,

$$c_1 = x_1 y_2', c_2 = x_1 y_3', \ldots, c_l = x_1 y_{l+1}'$$

for (mutually orthogonal) unit vectors $y_2', y_3', \ldots, y_{l+1}'$ in $\mathbb{R}^s$. Each of these vectors is perpendicular to y_1 because $x_1 y_1 = z$, being in $C_f^\perp$, is perpendicular to all collapse points. Retaining y_1, we now change the basis $\{y_1, y_2, \ldots, y_s\}$ of $\mathbb{R}^s$ to a new orthonormal basis $\{y_1, y_2', \ldots, y_{l+1}', y_{l+2}', \ldots, y_s'\}$, and replace B by the new set of mutually perpendicular unit vectors

$$B' = \{x_1 y_2', \ldots, x_1 y_{l+1}', x_1 y_{l+2}', \ldots, x_1 y_s'\}.$$

Clearly, each member of B' is still orthogonal to each member of the set A above. In particular, the entire C_f, spanned by $x_1 y_2', \ldots, x_1 y_{l+1}'$, is perpendicular to each member of the set A.

Finally we come to the exhibition of many great circular arc through z lying completely in Z. We simply make good use of the sets A and B'. For $k+1 \leq i \leq r$ there is a $\frac{1}{4}$-great circle γ_i of S^{n-1} running from $z = x_1 y_1$ to the vector $x_i y_1$ in A. This γ_i is just

$$x_1 y_1 \cos t + x_i y_1 \sin t = (x_1 \cos t + x_i \sin t) y_1, \; 0 \leq t \leq \frac{\pi}{2},$$

which firstly lies in V because both $x_1 \cos t + x_i \sin t$ and y_1 are of unit length. Secondly, γ_i in fact lies completely in Z, because both $x_1 y_1$ and $x_i y_1$ are perpendicular to C_f. Likewise for $l+2 \leq j \leq s$, we have a $\frac{1}{4}$-great circle $\tilde{\gamma}_j$ of S^{n-1} running from $z = x_1 y_1$ to the vector $x_1 y_j'$ in B'. This $\tilde{\gamma}_j$ is

$$x_1 y_1 \cos t + x_i y_j' \sin t = x_1 (y_1 \cos t + y_j' \sin t), \; 0 \leq t \leq \frac{\pi}{2}.$$

It also lies completely in Z because both x_1 and $y_1 \cos t + y_j' \sin t$ are unit vectors, and both $x_1 y_1$ and $x_1 y_j'$ are perpendicular to C_f, thanks to $l+2 \leq j \leq s$.

At the point $z \in Z$, the tangent vector to γ_i is a member of the set A, namely $x_i y_1$, and the tangent vector to $\tilde{\gamma}_j$ is a member of the set B', namely $x_1 y_j'$. Because of the orthogonal properties between A and B' mentioned above, the γ_i's and the $\tilde{\gamma}_j$'s are mutually perpendicular geodesics at z. Since $k+1 \leq i \leq r$ and $l+2 \leq j \leq s$, the total number of these geodesics is

$$(r-k)+(s-l-1)=r+s-k-l=d,$$

which is the dimension of Z as a submanifold in $S^{n-l-1} = C_f^{\perp} \cap S^{n-1}$. By Criterion 3.3, Z is a minimal submanifold S^{n-l-1}, as is to be proved.

We conclude this section by remarking that Theorem 3.1 does not exclude the case $l=0$, in which situation C is empty, and V is hypothesised to have regular points only. The theorem then simply asserts that V is a minimal submanifold of S^{n-1}.

4. Examples and remarks

In this section we shall give some illustrations to our Theorem 3.1.

Example 4.1. The "tensor product" multiplication $f : \mathbb{R}^r \times \mathbb{R}^s \longrightarrow \mathbb{R}^r \otimes \mathbb{R}^s = \mathbb{R}^{rs}$ given by $f(x, y) = x \otimes y$(or xy) is orthogonal, of type $[r, s, rs]$. This orthogonal multiplication has the feature that if x, x' and y, y' are independent in $\mathbb{R}^r$ and $\mathbb{R}^s$ respectively, then $x \otimes y, x' \otimes y, x \otimes y'$ and $x' \otimes y'$ are independent. Moreover, if $x' \perp x$ or $y' \perp y$, then $x' \otimes y' \perp x \otimes y$. It follows that the image of f consists of the so-called "rank-one tensors", i.e. tensors of the form $x \otimes y$, and factorisation is unique modulo the slight ambiguity that $x \otimes y = (\lambda x) \otimes (\frac{1}{\lambda} y)$, $\lambda \neq 0$. In particular, $V = f(S^{r-1} \times S^{s-1})$ has no collapse points, so $l = 0$. All points $z = x \otimes y$ in V are regular, with $\mathcal{L}F_z = \mathrm{span}\{x\}$, of dimension $k = 1$. The hypotheses in Theorem 3.1 are all satisfied, so one concludes that V is a minimal submanifold of dimension $r+s-2$, inside the sphere S^{rs-1}.

This is already a nontrivial example in that the topology of V is not quite simple. Note that V is doubly covered by $S^{r-1} \times S^{s-1}$ via the identification $(x, y) \sim (-x, -y)$. In [9] V is denoted by $S^{r-1} \times_T S^{s-1}$, and its cohomology ring is determined after some fair amount of work. Certainly, as a submanifold of S^{rs-1}, V is not totally geodesic.

It would be interesting to try to realise V as a minimal submanifold of spheres of lower (lowest ?) dimension.

Incidentally, the above discussions show that the product $\mathbb{R}P^{r-1} \times \mathbb{R}P^{s-1}$ of two real projective spaces is realisable as a minimal submanifold of $\mathbb{R}P^{rs-1}$. This is the famous Segre embedding.

Example 4.2. We skip the case of complex numbers and look at the quaternion analogue of Example 4.1. Denote the skew-field of quaternions by $\mathbb{H}$ and let $\mathbb{H}^r$ and $\mathbb{H}^s$ be, respectively, right and left $\mathbb{H}$-vector spaces. Let $\otimes_{\mathbb{H}}$ be abbreviated as $\otimes$. The tensor multiplication

$$f : \mathbb{H}^r \times \mathbb{H}^s \longrightarrow \mathbb{H}^r \otimes \mathbb{H}^s = \mathbb{R}^{4rs}$$

given by $f(x, y) = x \otimes y$ is orthogonal, of type $[4r, 4s, 4rs]$. Again, $V = f(S^{4r-1} \times S^{4s-1})$, contained in S^{4rs-1}, has no collapse point, and the hypotheses in Theorem 3.1 are satisfied with $l = 0$ and $k = 4$. Thus V is a $4r + 4s - 5$ dimensional submanifold of S^{4rs-1}. The product $S^{4r-1} \times S^{4s-1}$ smoothly fibers over V with S^3 fibers. Over $z = x \otimes y$ in V, the fiber can be identified with the unit sphere in the 4-dimensional space $\mathcal{L}F_z (= x\mathbb{H} = \mathbb{H}y)$. It would be interesting to know how V can be embedded minimally into spheres of lower dimension.

Example 4.3. The second exterior power gives a bilinear "multiplication"

$$g : \mathbb{R}^r \times \mathbb{R}^r \longrightarrow \mathbb{R} \oplus \Lambda^2(\mathbb{R}^r) = \mathbb{R}^n$$

where $n = 1 + \frac{1}{2}r(r-1)$. It is defined, for $x, y \in \mathbb{R}^r$, by

$$xy = g(x, y) = (\langle x, y\rangle, x \wedge y),$$

and is indeed an orthogonal multiplication, as verified below:

$$\begin{aligned} \|xy\|^2 &= (\langle x, y\rangle)^2 + \|x \wedge y\|^2 \\ &= (x_1y_1 + \cdots + x_ry_r)^2 + \sum_{1\le i<j\le r} (x_iy_j - x_jy_i)^2 \\ &= (x_1^2 + \cdots + x_r^2)(y_1^2 + \cdots + y_s^2) \\ &= \|x\|^2\|y\|^2. \end{aligned}$$

Here $x_1, \ldots, x_r$ and $y_1, \ldots, y_s$ are the Cartesian coordinates of x and y respectively. Put simply, g is the orthogonal multiplication associated to the "sum of squares identity" portraited by the third equality sign.

Since $x \wedge (\pm x) = 0$, and $x \wedge y = x' \wedge y'$ implies x, y, x', y' are coplanar in $\mathbb{R}^r$, it is easy to check that the space of collapse points C_g is precisely the direct summand $\mathbb{R}$ in $\mathbb{R} \oplus \Lambda^2(\mathbb{R}^r)$, while $C_g^\perp = \Lambda^2(\mathbb{R}^r)$. The variety $V = g(S^{r-1} \times S^{r-1})$ has two collapse points, $(1, 0)$ and $(-1, 0)$, while the rest of its points are regular, all having 2-dimensional left-factor spaces. So $k = 2$ and $l = 1$. Since $x' \perp x$ or $y' \perp y$ implies $x' \wedge y' \perp x \wedge y$, the orthogonal condition hypothesised in Theorem 3.1 is valid. Therefore $Z = C_g^\perp \cap V$ is a minimal submanifold in $S^{\frac{1}{2}r(r-1)-1}$, and its dimension is

$$d = r + r - k - l - 1 = 2r - 4.$$

Each point $z \in Z$ is of the form $(0, x \wedge y)$ where $x, y \in \mathbb{R}^r$ can be taken to be unital and orthogonal. This leads to a unique 2-plane in $\mathbb{R}^r$, namely $\mathcal{L}F_z = \text{span}\{x, y\}$, which is oriented by the ordered basis x, y. Thus, "by considering left-factor spaces", we can check easily that as a smooth manifold Z is the Grassmannian $\tilde{G}_2(\mathbb{R}^r)$ of oriented 2-planes in $\mathbb{R}^r$, now embedded minimally into $S^{\frac{1}{2}r(r-1)-1}$. This embedding is none other than a lifting of the Plücker embedding of $G_2(\mathbb{R}^r)$, the Grassmannian manifold of unoriented 2-planes, into an appropriate real projective space. This Plücker embedding, therefore, is also minimal.

The algebraic variety V in this example is, by Lemma 2.5, the joint $C * Z$, or $S^0 * Z$, which is just the suspension of $\tilde{G}_2(\mathbb{R}^r)$.

Example 4.4. The situation in Example 4.3 has a complex analogue. Give $\mathbb{C}^r$ the usual Hermitian product $\langle z, w\rangle_{\mathbb{C}} = \sum \bar{z}_i w_i$. The $\mathbb{R}$-bilinear map

$$h : \mathbb{C}^r \times \mathbb{C}^r \longrightarrow \mathbb{C} \oplus \Lambda^2(\mathbb{C}^r) = \mathbb{C}^n$$

with $n = 1 + \frac{1}{2}r(r-1)$, given by

$$h(z, w) = \left(\sum_{i=1}^{r} \bar{z}_i w_i, z \wedge w\right)$$

is an orthogonal multiplication of type $[2r, 2r, 2 + r(r-1)]$. Here $z = (z_1, \ldots, z_r)$, $w = (w_1, \ldots, w_r)$ and the wedge product pairing into $\Lambda^2(\mathbb{C}^n)$ is taken over $\mathbb{C}$. Quite analogous to Example 4.3, we get a minimal $(4r-7)$ dimensional submanifold Z inside the sphere $S^{r(r-1)-1} = S^{2n-3}$.

From another point of view this minimal submanifold is of course closely related to the Plücker embedding p of the complex Grassmannian $G_2(\mathbb{C}^r)$ into complex projective space $\mathbb{C}P^{\frac{1}{2}r(r-1)-1} = \mathbb{C}P^{n-2}$. Because p is a holomorphic map, it is certainly harmonic, so $G_2(\mathbb{C}^r)$ is a minimal submanifold of $\mathbb{C}P^{n-2}$. The canonical line bundle H of $\mathbb{C}P^{n-2}$, when restricted to $G_2(\mathbb{C}^r)$, is simply $\Lambda^2(\gamma)$, where γ is the tautological complex 2-plane bundle over $G_2(\mathbb{C}^r)$. Thus P induces an inclusion of the circle bundle of $\Lambda^2(\gamma)$ into that of H, in an obvious way. Clearly, our Z is the total space of the first circle bundle, and S^{2n-3} that of the second. In this sense our conclusion that Z is a minimal submanifold of S^{2n-3} isn't too surprising: it is, somehow, a consequence of holomorphicity.

In conclusion we like to point out that the results in this paper lead to at least two interesting questions. One is to exhaust all those orthogonal multiplications that satisfy the hypotheses of Theorem 3.1, and to investigate each minimal submanifold so obtained. The other is to further pursue the special class of submanifolds of spheres which are minimal because they satisfy Criterion 3.3. It seems that this kind of study has not been given in the differential geometry literature.

On the other hand it is conceivable that Theorem 3.1 could be improved in some way, and more submanifolds of spheres could be thus obtained. Further work remains to be done.

As a side remark the suggestive connection between orthogonal multiplications and Plücker embeddings, as given in our Examples, certainly leads to the natural question: is the general Plücker embedding

$$p : G_k(\mathbb{R}^m) \longrightarrow \mathbb{R}P^{n-1}, \; n = \binom{m}{k}$$

of the Grassmann space of real k-planes in $\mathbb{R}^m$ into the appropriate real projective space, a harmonic map with respect to the conventional Riemannian metric of these spaces? The answer is yes, as is probably known already to algebraic and differential geometers. But the following short proof might have some pertinent interest: one can first pass to the embedding $\tilde{p} : \tilde{G}_k(\mathbb{R}^m) \longrightarrow S^{\binom{m}{k}-1}$ of oriented Grassmannians into the appropriate spheres, then use Criterion 2.2 to show that $\tilde{p}$ is a harmonic map. The steps are quite straightforward, if one uses the description of the tangent bundle of $\tilde{G}_k(\mathbb{R}^m)$ furnished in [6], Corollary 1.2. Harmonicity of p will then follow, but we shall omit further details.

References

[1] P. Baird and Y. L. Ou, Harmonic maps and morphisms from multilinear norm-preserving mappings, *Internat. J. Math.*, **8(2)** (1997) 187–211.

[2] W. Ding, Harmonic Hopf constructions between spheres, *Internat. J. Math.*, **5** (1994) 849–860.

[3] J. Eells and L. Lemaire, A report on harmonic maps, *Bull. London Math. Soc.*, **10(1)** (1978) 1–68.

[4] J. Eells and A. Ratto, *Harmonic maps and minimal immersions with symmetries*, vol. 130 of *Annals of Mathematics Studies*, Princeton University Press, Princeton, NJ (1993). Methods of ordinary differential equations applied to elliptic variational problems.

[5] A. Hurwitz, Über die komposition der quadratischen formen von beliebig vielen variablen, *Nach. Akad. Wiss. Gött. II, Math. Phys.*, **K1** (1898) 309–316.

[6] K. Y. Lam, A formula for the tangent bundle of flag manifolds and related manifolds, *Trans. Amer. Math. Soc.*, **213** (1975) 305–314.

[7] K. Y. Lam, Non-singular bilinear maps and stable homotopy classes of spheres, *Math. Proc. Cambridge Philos. Soc.*, **82(3)** (1977) 419–425.

[8] K. Y. Lam, Some new results on composition of quadratic forms, *Invent. Math.*, **79(3)** (1985) 467–474.

[9] K. Y. Lam, Borsuk-Ulam type theorems and systems of bilinear equations. In *Geometry from the Pacific Rim (Singapore, 1994)*, pages 183–194. de Gruyter, Berlin (1997).

[10] K. Y. Lam and P. Y. H. Yiu, Geometry of normed bilinear maps and the 16-square problem, *Math. Ann.*, **284(3)** (1989) 437–447.

[11] B. Lawson, *Lectures on Minimal Submanifolds*, vol. 1. Berkeley, CA 94704 USA (1980).

[12] C. K. Peng and Z. Z. Tang, On representing homotopy classes of spheres by harmonic maps, *Topology*, **36(4)** (1997) 867–879.

[13] J. Radon, Linear scharen orthogonaler matrizen, *Abh. Math. Sem. Univ. Hamburg*, **1** (1922) 1–148.

[14] A. Ratto, Harmonic maps of spheres and the Hopf construction, *Topology*, **28(3)** (1989) 379–388.

[15] D. B. Shapiro, Compositions of quadratic forms, vol. 33 of *de Gruyter Expositions in Mathematics*, Walter de Gruyter & Co., Berlin (2000).

[16] R. T. Smith, Harmonic mappings of spheres, *Amer. J. Math.*, **97** (1975) 364–385.

[17] P. Y. H. Yiu, Sums of squares formulae with integer coefficients, *Canad. Math. Bull.*, **30(3)** (1987) 318–324.

Perspectives in Geometry and Topology No. 11, 2010, pp. 29–47.

Lectures on Topological Methods in the study of Bilinear Forms

Kee Yuen Lam

Department of Mathematics, University of British Columbia, Vancouver, B.C., V6T 1Z2 Canada
e-mail: keeyuen.lam@gmail.com

Abstract. The study of many problems in algebra can benefit greatly by employing topological methods such as homotopy theory and cohomology theories. In these lectures we provide a comprehensive account on the usefulness of such an approach in studying systems of bilinear forms which arise frequently in algebra and in geometry.

Keywords. Equation solving, bilinear forms, Hurwitz-Radon function, projective spaces, the Stiefel-Hopf condition, cohomology theories.

1. Lecture

1.1 *Bilinear forms and systems of homogeneous bilinear equations*

Let $x_1, \dots, x_r$ and $y_1, \dots, y_s$ be two sets of independent variables. By a bilinear form in these variables we mean an expression

$$\phi = \sum_{\substack{1 \le p \le r \\ 1 \le q \le s}} a_{pq} x_p y_q .$$

Throughout these lectures, the coefficients a_{pq} shall be real numbers although they could be elements in a field $\mathbf{F}$ in more general circumstances. Also, we will allow the variables x_p and y_q to assume real values only. One may regard ϕ as an element in a polynomial ring $\mathbb{R}[x_1, \dots, x_r; y_1, \dots, y_s]$. But frequently the study of ϕ is done via expressing

$$\phi = [x_1, \dots, x_r] A [y_1, \dots, y_s]^t$$

where $A = [a_{pq}]$ is an $r \times s$ matrix with entries in $\mathbb{R}$. When there are several bilinear forms taken together as a family:

$$\phi^j = \sum_{\substack{1 \le p \le r \\ 1 \le q \le s}} a^j_{pq} x_p y_q, \quad 1 \le j \le n$$

involving n matrices $A(j)$, $(1 \le j \le n)$ of the same size $r \times s$, the study becomes nontrivial, and leads to interesting questions. To motivate the main question in this workshop, consider an arbitrary system (1) of homogeneous bilinear equations of type (r, s, n):

$$\begin{array}{rcl} \phi^1(x_1, \ldots, x_r; y_1, \ldots, y_s) & = & 0 \\ \cdots & \cdots & \cdots \\ \phi^n(x_1, \ldots, x_r; y_1, \ldots, y_s) & = & 0 \end{array} \tag{1}$$

Obviously it admits trivial solutions, namely those corresponding to either all the r unknowns $x_p = 0$ or all the s unknowns $y_q = 0$ (or both). The question is: under what circumstances will nontrivial (real-valued) solutions exist for all possible values of the coefficients? The analogous question for arbitrary homogeneous linear systems

$$\begin{array}{rcl} b_{11}z_1 + \cdots + b_{1m}z_m & = & 0 \\ \cdots & \cdots & \cdots \\ b_{n1}z_1 + \cdots + b_{nm}z_m & = & 0 \end{array} \tag{2}$$

with m unknowns and n equations, is well understood. For such systems, the condition

$$m > n \tag{3}$$

guarantees the existence of nontrivial real solutions to $z_1, \ldots, z_m$, irrespective of the values of the coefficients b_{ij}. In other words, nontrivial solvability of (2) over $\mathbb{R}$ is assured if the number of variables exceeds the number of equations. Returning to the system (2), it is now an interesting question to impose suitable numerical conditions on r, s and n, similar to the one given in (3), which can assure the existence of nontrivial solutions for $x_1, \ldots, x_r; y_1, \ldots, y_s$ over $\mathbb{R}$, irrespective of the values of the coefficients. We shall address this question in the subsequent sections.

Exercise 1.

(a) Show algebraically that any system of 3 homogeneous bilinear equations in the unknowns $x_1, x_2; y_1, y_2, y_3$ is always nontrivially solvable.
(b) Write down a system of 4 homogeneous bilinear equations in the same set of unknowns which has trivial solutions only.
(c) Prove by topological methods that any system of 6 homogeneous bilinear equations in unknowns $x_1, x_2, x_3; y_1, y_2, y_3, y_4, y_5$ is always nontrivially solvable.

1.2 *Nonsingular bilinear maps; classical examples*

The question raised in §1.1 can be reformulated in another suitable way. Consider the bilinear map

$$\Phi : \mathbb{R}^r \times \mathbb{R}^s \to \mathbb{R}^n$$

given by

$$\Phi(x, y) = (\Phi^1(x, y), \ldots, \Phi^n(x, y)) \in \mathbb{R}^n$$

where x stands for $(x_1, \ldots, x_r)$ and y for $(y_1, \ldots, y_s)$. We say that Φ is *nonsingular* if $\Phi(x, y) = 0$ implies $x = 0$ or $y = 0$. Our reformulated question is:

$$\boxed{\text{Determine all triples } (r, s, n) \text{ for which there exists a nonsingular bilinear map } \Phi : \mathbb{R}^r \times \mathbb{R}^s \to \mathbb{R}^n} \tag{4}$$

These are, of course, precisely the triples for which a system (1) of type (r, s, n) may fail to have nontrivial solutions.

Examples of nonsingular bilinear maps abound in elementary mathematics:

(i) If F denotes the (not necessarily commutative) field of real, complex or quaternion numbers $\mathbb{R}$, $\mathbb{C}$, or $\mathbb{H}$, their multiplication

$$\mu : F \times F \to F$$

is a nonsingular bilinear map of type (n, n, n), $n = 1, 2$ or 4.
(ii) An *octonion number*[1] is, by definition, a pair of quaternions $(a_1, a_2), a_i \in \mathbb{H}$. They can be multiplied according to the formula

$$(a_1, a_2) \times (b_1, b_2) = (a_1 b_1 - \bar{b}_2 a_2, b_2 a_1 + a_2 \bar{b}_1). \tag{5}$$

This is the same formula used to define quaternion multiplication when a quaternion is regarded as an ordered pair of complex numbers. The octonion multiplication $\mu : \mathbb{O} \times \mathbb{O} \to \mathbb{O}$ is another example of a nonsingular bilinear map, of type $(8, 8, 8)$. It turns $\mathbb{O}$ into a "division algebra" with neither commutativity nor associativity, and it completes the list of division algebras

$$\mathbb{R} \subset \mathbb{C} \subset \mathbb{H} \subset \mathbb{O}$$

that can exist over $\mathbb{R}$. (See [7] for more information). Note, in particular, the useful property that $\mathbb{R}$ is the center of $\mathbb{O}$, and that any sub-algebra of $\mathbb{O}$ with 2 generators over $\mathbb{R}$ is associative.

Exercise 2. Show that Octonion multiplication still enjoys the norm property, namely that

$$|\mu(a, b)| = |a||b|$$

where $|\cdot|$ denotes the Euclidean norm.

Exercise 3. Introduce a multiplication of pairs of octonions by the same formula as in (5), thereby getting a bilinear map

$$\Psi : \mathbb{R}^{16} \times \mathbb{R}^{16} \to \mathbb{R}^{16}.$$

[1] Also known as *Cayley number.*

(a) Show that Ψ isn't nonsingular;
(b) Find subspaces U, V, W of $\mathbb{R}^{16}$ such that Ψ restricts to a nonsingular bilinear map $U \times V \to W$. What kind of types are obtained by this restriction process?

(iii) We shall use a special case of Ex. 2(b). If U denotes the subspace of pairs (a_1, a_2) of octonions where a_2 is real, (i.e., $a_2 \in \mathbb{R} \subset \mathbb{O}$), then $U \subset \mathbb{R}^{16}$; $U \cong \mathbb{R}^9$ and it can be checked that the composite map Φ :

$$\mathbb{R}^9 \times \mathbb{R}^{16} \to \mathbb{R}^{16} \times \mathbb{R}^{16} \xrightarrow{\Psi} \mathbb{R}^{16}$$

is bilinear. Furthermore the norm condition

$$|\Phi(x, y)| = |x||y|, \quad \forall\, x \in \mathbb{R}^9, y \in \mathbb{R}^{16}$$

is still valid, so Φ is certainly nonsingular. Its type is (9, 16, 16). The standard basis of $\mathbb{R}^9$ induces, via Φ, orthogonal transformations $\gamma_0, \gamma_1, \ldots, \gamma_8$ of $\mathbb{R}^{16}$, with $\gamma_0 = I$. It can be checked that the norm condition is equivalent to

$$\begin{aligned} \gamma_i^2 &= -I \qquad \text{for } 1 \leq i \leq 8 \\ \gamma_{ij} + \gamma_{ji} &= 0 \ \ \text{for } 1 \leq i, j \leq 8, i \neq j \end{aligned} \tag{6}$$

(iv) The special example in (iii) can be used to generate many more examples of nonsingular bilinear maps with the norm property. If

$$F : \mathbb{R}^r \times \mathbb{R}^n \to \mathbb{R}^n$$

satisfies $|F(x, y)| = |x||y|$ for all $x \in \mathbb{R}^r$, $y \in \mathbb{R}^s$, we can let the standard basis of $\mathbb{R}^r$ induce, via F, orthogonal transformations, $\theta_1, \ldots, \theta_r$ of $\mathbb{R}^n$ to itself. By suitable choice of coordinates in the range, it is possible to arrange that $\theta_1 = I$. If so, one would have the relations

$$\begin{aligned} \theta_i^2 &= -I \qquad \text{for } 2 \leq i \leq r \\ \theta_i\theta_j + \theta_j\theta_i &= 0 \ \ \text{for } 2 \leq i, j \leq r, i \neq j \end{aligned} \tag{7}$$

just as in (6).
One can now construct a set of $r + 8$ orthogonal transformations acting on $\mathbb{R}^{16n} = \mathbb{R}^n \otimes \mathbb{R}^{16}$. They are

$$\{I \otimes I, \theta_2 \otimes \Gamma, \ldots, \theta_r \otimes \Gamma, I \otimes \gamma_1, \ldots, I \otimes \gamma_8\}$$

where Γ denotes the product of $\gamma_1, \ldots, \gamma_8$. It is straightforward to check that $\Gamma^2 = I$ and that if these transformations are renamed as $\{(I =)\omega_1, \omega_2 \ldots, \omega_{r+7}\}$ then they satisfy the same type of relations as the θ's in (7)). Consequently, we can build a new bilinear map

$$\mathbb{R}^{r+8} \times \mathbb{R}^{16n} \to \mathbb{R}^{16n}$$

by letting the transformations on $\mathbb{R}^{16n}$ induced by the standard basis of $\mathbb{R}^{r+8}$ to be precisely the ω's. This new map again has the norm property, and is certainly nonsingular.

Exercise 4 (Hurwitz-Radon maps). Show that if $n = 2^{a+4b}(2c+1)$ where $0 \leq a \leq 3$, then there is a bilinear map

$$h : \mathbb{R}^{\rho(n)} \times \mathbb{R}^n \to \mathbb{R}^n \tag{8}$$

with the norm property, where $\rho(n) = 2^a + 8b$ is the so-called Hurwitz-Radon function. Historically, this family of maps had been thoroughly investigated by algebraists and topologists.

1.3 *Nonsingular bilinear maps, additional examples*

Let a vector $(a_o, \ldots, a_{r-1})$ of $\mathbb{R}^r$ be identified with the real-coefficient polynomial $\sum_{i=0}^{r-1} a_i T^i$ in the variable T. Then multiplication of polynomials produces a nonsingular bilinear map

$$P : \mathbb{R}^r \times \mathbb{R}^s \to \mathbb{R}^{r+s-1} \tag{9}$$

given by

$$P((a_0, \ldots, a_{r-1}), (b_0, \ldots, b_{s-1})) = (c_0, c_1, \ldots, c_{r+s-2}), \tag{10}$$

with $c_k = \sum_{i+j=k} a_i b_j$. Note that P does not enjoy the norm property, unless $r = 1$ or $s = 1$.

If, in the above discussion, $\mathbb{R}$ is replaced by the division algebra $\mathbf{F} = \mathbb{C}, \mathbb{H}$ or $\mathbb{O}$ regarded as $\mathbb{R}^2, \mathbb{R}^4$ or $\mathbb{R}^8$ respectively, the same formula (10) will produce examples

$$\tilde{P} : \mathbb{R}^{dr} \times \mathbb{R}^{ds} \to \mathbb{R}^{d(r+s-1)} \tag{11}$$

with $d = \dim_{\mathbb{R}sm} \mathbf{F} = 2, 4$, or 8. $\tilde{P}$ is nonsingular bilinear over $\mathbb{R}$. Furthermore, some twisting is possible to produce other examples. For instance, in the definition of $\tilde{P}$, $c_2 = a_0b_2 + a_1b_1 + a_2b_0$ can be replaced by $c_2 = a_0b_2 + \bar{b}_1a_1 + a_2b_0$ where $\bar{b}_1$ means the conjugate of b_1 in the respective division algebra.

Exercise 5. Formulate and prove a fundamental theorem of algebra for polynomials with quaternion or octonion coefficients.

Exercise 6 (Study of linear polynomials over the octonions $\mathbb{O}$).

(a) Show that the nonsingular bilinear map

$$P : \mathbb{R}^{16} \times \mathbb{R}^{16} \to \mathbb{R}^{24}$$

defined by the multiplication of linear polynomials as in (10) is onto.

(b) Find a suitable twisting: $\tilde{P} : \mathbb{R}^{16} \times \mathbb{R}^{16} \to \mathbb{R}^{24}$ of P that is no longer onto.

(c) Prove that there exists a nonsingular bilinear map of type (16, 16, 23).

(d) Must a system of 23 homogeneous bilinear equations in the unknowns $x_1, \ldots, x_{16}$; $y_1, \ldots, y_{16}$ be nontrivially solvable?

1.4 *The Stiefel-Hopf condition [4]*

Continuing with the discussions of Exercise 6(d), one may wonder whether there can exist a nonsingular bilinear map of type (16, 16, 22). The nontrivial answer turns out to be NO, as we shall subsequently see. This means that Exercise 5(d) would have an opposite conclusion if the 23 there is replaced by 22.

To state the Stiefel-Hopf condition which limits the possible types (r, s, n) of nonsingular bilinear maps, a numerical function $r \circ s$ is required. Let $\mathbb{Z}_2$ be the field of 2 elements and let $\Lambda = \mathbb{Z}_2[u, v]/(u^r, v^s)$ be the quotient ring of the polynomial ring $\mathbb{Z}_2[u, v]$ by the ideal generated by u^r and v^s. By definition, $r \circ s$ is the height of the element $(u + v)$ in Λ, that is, the positive integer h such that $(u + v)^h = 0$ but $(u + v)^{h-1} \neq 0$. Quite obviously,

$$\operatorname{Max}\{r, s\} \leq r \circ s \leq r + s - 1.$$

Theorem 1 (Stiefel-Hopf). *If there exists a nonsingular bilinear map* $\Phi : \mathbb{R}^r \times \mathbb{R}^s \to \mathbb{R}^n$ *then* $n \geq r \circ s$.

To see the relevance of this theorem to our main question, note that it implies nontrivial solvability of the system (1) whenever $n < r \circ s$, or whenever $r + s > n + (r + s - r \circ s)$. Thus for arbitrary homogeneous bilinear systems, nontrivial solvability over $\mathbb{R}$ is guaranteed by the condition that the number $r + s$ of unknowns exceed the number n of equations by a gap wider than $r + s - r \circ s$. This is exactly in the same spirit as condition (3) is for homogeneous linear system.

Exercise 7.

(a) Evaluate $3 \circ 5$ from definition.

(b) Establish the following properties of the function $r \circ s$:

$$r \circ s = s \circ r, \quad 1 \circ s = s \text{ and } 2^t \circ 2^t = 2^t;$$

$$r \circ s = \begin{cases} 2^t & \text{if } 2^{t-1} < r, s \leq 2^t, \\ 2^{t-1} + r \circ (s - 2^t) & \text{if } r \leq 2^{t-1} < s. \end{cases}$$

(c) Use (b) to quickly evaluate $27 \circ 99$.

(d) Generalizing (b), prove that if $r \leq 2^{t-1} \leq M < s$ where 2^{t-1} divides M, then $r \circ s = M + r \circ (s - M)$. Use this to evaluate $27 \circ 99$ again.

2. Lecture

2.1 *Motivation from Material Science*

In material science, between macrostructures and atomic structures, there is a level of microstructure such as found in some alloys. Consider a simple microstructure

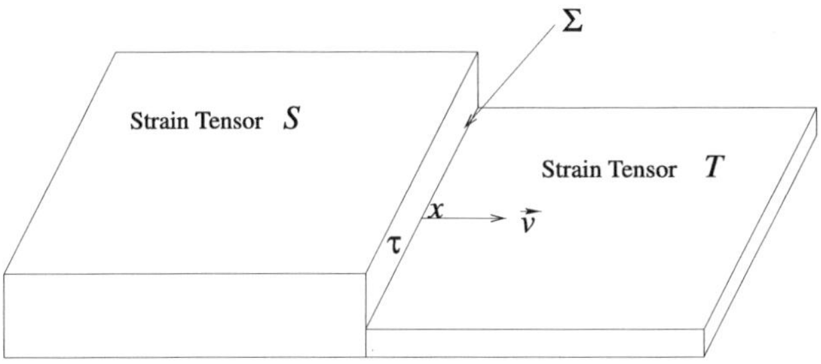

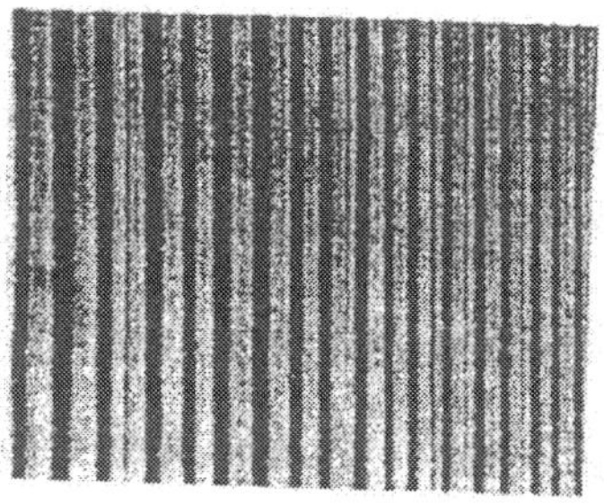

consisting of two distinct kinds of materials, say in alternate layers, interfacing at surface Σ. At a point $x \in \Sigma$ there are 2 distinct strain tensors S and T of type $(3, 3)$, i.e., $S, T \in \mathbb{R}^3 \otimes \mathbb{R}^3$ corresponding to the two materials. Let

$$\tau = \text{the tangent plane to } \Sigma \text{ at } x$$

$$\vec{v} = \text{a unit vector normal to } \Sigma \text{ at } x.$$

According to the direct sum decomposition

$$\tau \oplus \mathbb{R}\vec{v} = \mathbb{R}^3$$

where $\mathbb{R}\vec{v}$ denotes the 1-dimensional vector space generated by $\vec{v}$, one has the decomposition

$$\mathbb{R}^3 \otimes \mathbb{R}^3 = (\tau \oplus \mathbb{R}\vec{v}) \otimes \mathbb{R}^3 = \tau \otimes \mathbb{R}^3 \oplus (\mathbb{R}\vec{v}) \otimes \mathbb{R}^3$$

The law of microstructure says that for S, T to be physically feasible they should have the same component along $\tau \otimes \mathbb{R}^3$ under this decomposition. Thus

$$S - T = \vec{v} \otimes \vec{\omega}, \text{ for some } \vec{\omega} \in \mathbb{R}^3.$$

In other words span $\{S, T\}$, as a subspace of $\mathbb{R}^3 \otimes \mathbb{R}^3$, must contain a tensor of rank 1. Consequently the composite bilinear map Φ:

$$\mathbb{R}^3 \times \mathbb{R}^3 \xrightarrow{\text{canonical}} \mathbb{R}^3 \otimes \mathbb{R}^3 \xrightarrow{\text{quotient}} \mathbb{R}^3 \otimes \mathbb{R}^3/\text{span}\{S, T\}$$

satisfies $\Phi(\vec{v}, \vec{\omega}) = 0$. This shows that singularity of bilinear maps is a built-in feature of microstructure theory.

The above discussion also leads to the following

Observation. Let $\Phi : \mathbb{R}^r \times \mathbb{R}^s \to \mathbb{R}^s$ be a bilinear map inducing the linear map $L : \mathbb{R}^r \otimes \mathbb{R}^s \to \mathbb{R}^n$ characterized by $L(\vec{u} \otimes \vec{v}) = \Phi(\vec{u}, \vec{v})$. Then Φ is nonsingular if and only if K, the kernel of L, is a subspace of $\mathbb{R}^r \otimes \mathbb{R}^s$ which avoids all tensors of rank 1, i.e., contains no tensor of the type $\vec{u} \otimes \vec{v}$ with $\vec{u} \in \mathbb{R}^r, \vec{v} \in \mathbb{R}^s, \vec{u} \neq 0 \neq \vec{v}$. Thus the search for all nonsingular bilinear maps with domain $\mathbb{R}^r \times \mathbb{R}^s$ is equivalent to the determination of all linear subspaces K of $\mathbb{R}^r \otimes \mathbb{R}^s$ that do not contain nonzero tensors of the form $\vec{u} \otimes \vec{v}$, which are the so-called rank one tensors.

2.2 *Behrend's proof of the Stiefel-Hopf condition [3]*

In the special case $n = r+s-2$, the Stiefel-Hopf theorem translates into the following:

Proposition 1. *If there exists a nonsingular bilinear map*

$$\Phi : \mathbb{R}^r \times \mathbb{R}^s \to \mathbb{R}^{r+s-2}$$

then $r + s - 2 \geq r \circ s$.

Remark 1. The concluding inequality means that in the quotient ring $\Lambda = \mathbb{Z}_2[u, v]/(u^r, v^r)$ introduced in 1.4, $(u + v)^{r+s-2} = 0$. Binomially, the left side is $\binom{r+s-2}{r-1}u^{r-1}v^{s-1}$ so the proposition will be proved by showing that $\binom{r+s-2}{r-1} \equiv 0 \pmod 2$. What Behrend did was to assume that $\binom{r+s-2}{r-1}$ were odd, and deduce from it that Φ must be singular. For this purpose, write $n = r + s - 2$ for brevity and form the diagram

$$\begin{array}{ccccc}
 & & \mathbb{R}^r \times \mathbb{R}^s & \xrightarrow{\Phi} & \mathbb{R}^n \\
 & & \downarrow{\scriptstyle \Delta} & \nearrow{\scriptstyle L} & \\
\text{Ker } L & \xrightarrow{j} & \mathbb{R}^r \otimes \mathbb{R}^s & &
\end{array}$$

where Δ is the canonical map, L is the linear map induced by Φ and j is the inclusion map of the kernel of L into $\mathbb{R}^r \otimes \mathbb{R}^s$. By taking the projectivization of various linear spaces appearing in this diagram, one gets

$$\begin{array}{ccccc}
 & & P(\mathbb{R}^r) \times P(\mathbb{R}^s) & \xrightarrow{P\Phi} & P(\mathbb{R}^n) \\
 & & \downarrow{\scriptstyle P\Delta} & \nearrow{\scriptstyle PL} & \\
P(\text{Ker } L) & \xrightarrow{Pj} & P(\mathbb{R}^r \otimes \mathbb{R}^s) & &
\end{array}$$

This diagram shows that, inside the projective space $P(\mathbb{R}^r \otimes \mathbb{R}^s)$ of dimensions $rs - 1$, there are two mod2 homology cycles, geometrically represented by embedded submanifolds of dimensions $(r-1)+(s-1)$ and $k-1$ respectively, via $P\Delta$ and Pj. Here $k = \dim(\text{Ker } L) \geq rs - n$. In how many points must these two geometrical cycles intersect? To find out, Behrend first replaced Ker L by a certain subspace J of $\mathbb{R}^r \otimes \mathbb{R}^s$, defined in the following manner: Pick nonzero real numbers $a_1, \ldots, a_r; b_1, \ldots, b_s$ such that the numbers $a_i b_j$ are mutually distinct. Write down matrices

$$U = \begin{pmatrix} a_1 & \cdots & a_r \\ a_1^2 & \cdots & a_r^2 \\ \cdots & \cdots & \cdots \\ a_1^n & \cdots & a_r^n \end{pmatrix}; \quad V = \begin{pmatrix} b_1 & \cdots & b_s \\ b_1^2 & \cdots & b_s^2 \\ \cdots & \cdots & \cdots \\ b_1^n & \cdots & b_s^n \end{pmatrix}$$

and

$$C = \begin{pmatrix} a_1b_1 & a_1b_2 & \cdots & a_rb_s \\ (a_1b_1)^2 & (a_1b_2)^2 & \cdots & (a_rb_s)^2 \\ \cdots & \cdots & \cdots & \cdots \\ (a_1b_1)^n & (a_1b_2)^n & \cdots & (a_rb_s)^n \end{pmatrix}$$

of sizes $n \times r$, $n \times s$ and $n \times rs$ respectively. Then with reference to the standard basis of $\mathbb{R}^r \otimes \mathbb{R}^s$, J is the subspace consisting of vectors whose rs components W, arranged in matrix format, satisfy

$$\sum_{i,j} a_i^\mu w_{ij} b_j^\mu = 0, \quad \mu = 1, \dots, n.$$

In other words, J is the null space of the matrix C, and in yet other words J consists precisely of those coordinate vectors W such that UWV^t has zero diagonal. Here the superscript ^{t} stands for transpose. Consider an intersection point $[W]$ of $P(\mathbb{R}^r) \times P(\mathbb{R}^s)$ with $P(J)$. It can be written as $[W] = [XY^t]$ where X is $r \times 1$, Y is $s \times 1$, $X \neq 0 \neq Y$. It must also be such that the $n \times n$ matrix

$$U(XY^t)V^t = (UX)(VY)^t \tag{12}$$

has zero diagonal. Because all $r \times r$ sub-matrices of U have rank r, and all $s \times s$ sub-matrices of V have rank s, the $n \times 1$ column UX has no more than $r - 1$ zeros, and the $1 \times n$ row $(VY)^t$ has no more than $s - 1$ zeros. Remembering that $n = r + s - 2$ is the number of diagonal entries in (12), one concludes that UX has $r - 1$ zero components, and VY has $(s - 1)$, situated in positions complementary to those of UX. Therefore the number of intersection points of $P(\mathbb{R}^r) \times P(\mathbb{R}^s)$ with $P(J)$ is equal to $\binom{r+s-2}{r-1}$, corresponding to the various ways of placing $r - 1$ zeros in a $n \times 1$ column. Now, because rank $C = n$, the dimension of J is $rs - n$. Should $\binom{r+s-2}{r-1}$ be odd, the number of intersection points will be nonzero mod 2, and things will remain so as J moves continuously through equi-dimensional subspaces of $\mathbb{R}^r \otimes \mathbb{R}^s$ until it becomes a linear subspace of $\ker L$, proper or improper. This is because "*in real algebraic geometry, the count of intersection points always change by an even amount*" (see picture below):

In particular, the oddness of $\binom{r+s-2}{r-1}$ would entail at least one point of intersection between $P(\mathbb{R}^r) \times P(\mathbb{R}^s)$ with $P(\ker L)$. This means that there are nonzero vectors $u \in \mathbb{R}^r, v \in \mathbb{R}^s$ with $\Phi(u, v) = L(u \times v) = 0$, so that Φ is singular, contrary to hypothesis. Q.E.D.

Exercise 8. By considering various restrictions of nonsingular bilinear maps, show that Behrend's Proposition above implies the full Stiefel-Hopf theorem.

Exercise 9. For the purpose of rigor, the proof above should include a check that the submanifolds $P(\mathbb{R}^r) \times P(\mathbb{R}^s)$ and $P(J)$ intersect transversally. Supply this check.

Exercise 10. The *italicized assertion* in the above proof should be made more precise and transparent. Typically, this can be done by replacing intersection of cycles by cup products of cocycles. Supply a proof of the Proposition by considering the induced ring homomorphism Δ^* in cohomology. (This is a classical argument, given by H. Hopf in 1941; see also Lecture 3 below).

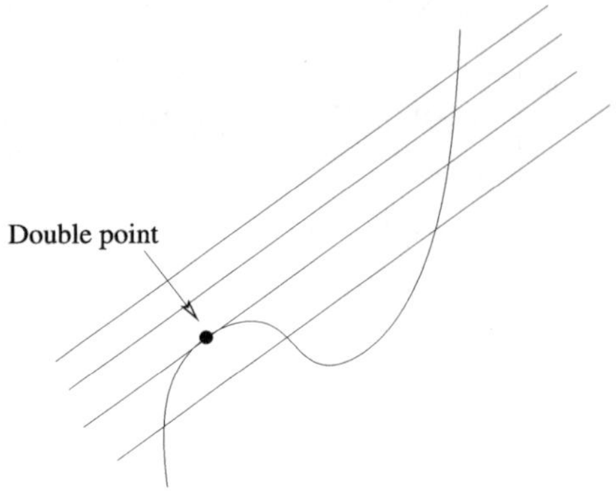

2.3 *Atiyah's necessary condition for the existence of a nonsingular bilinear map [2]*

Atiyah used K-theory to obtain another necessary condition for the existence of a nonsingular bilinear map Φ of type (r, s, n). The idea is to employ arguments similar to that of Behrend and Hopf, but in the context of K-cohomology which is based on considerations of vector bundles. Here is his argument in a nut-shell. We have already encountered the real projective space $P(\mathbb{R}^r)$ which we now re-denote by $\mathbb{R}P^{r-1}$ to emphasize that it is a differentiable manifold of dimension $r-1$. Let ξ_{r-1} and ϵ denote, respectively, the canonical and the trivial line bundle over $\mathbb{R}P^{r-1}$. Then one has:

Lemma 1. *If there exists a nonsingular bilinear map Φ of type (r, s, n), then $n\xi_{r-1}$ contains an s-dimensional trivial sub-bundle, where*

$$n\xi_{r-1} = \xi_{r-1} \oplus \cdots \oplus \xi_{r-1}$$

is the n-fold Whitney sum of ξ_{r-1} with itself.

The proof is quite simple. Recall that $\mathbb{R}P^{r-1}$ is obtainable from the sphere S^{r-1} by identifying each point x with its antipodal point $-x$. The total space of $n\xi_{r-1}$ is obtained from $S^{r-1} \times \mathbb{R}^n$ by identifying each (x, v) with $(-x, -v)$. Let us denote it by $S^{r-1} \times_T \mathbb{R}^n$. Clearly, it contains the total space of $s\epsilon$ as a vector sub-bundle, the inclusion map being given by

$$i : \mathbb{R}P^{r-1} \times \mathbb{R}^s \to S^{r-1} \times_T \mathbb{R}^n.$$

$$([x], v) \mapsto [x, \Phi(x, v)]. \qquad \text{Q.E.D.}$$

The above lemma allows us to write down an equality of vector bundles

$$n\xi = \eta \oplus s\epsilon \tag{13}$$

where ξ abbreviates for ξ_{r-1}, and η is the vector sub-bundle inside $n\xi$ complementary to $s\epsilon$. We now bring in the exterior power functors λ^k in linear algebra. These associate, functorially, a real vector space $\lambda^k(V)$ to any q-dimensional real vector space

V, with $\dim_{\mathbb{R}sm} \lambda^k(V) = \binom{q}{k}$. In particular, $\lambda^0(V) = \mathbb{R}$, $\lambda^1(V) = V$ and $\lambda^k(V) = 0$ if $k > \dim(V)$, while the Whitney formula

$$\lambda^k(V \oplus W) \approx \oplus_{0 \leq j \leq k} \lambda^j(V) \otimes \lambda^{k-j}(W) \tag{14}$$

is always valid. If we adopt as a formal expression

$$\lambda_t(V) = \lambda^0(V) + \lambda^1(V)t + \lambda^2(V)t^2 + \cdots$$

and view it as a formal power series in t with coefficients being vector spaces (and later, vector bundles), we can expand (14) into

$$\lambda_t(V \oplus W) = \lambda_t(V)\lambda_t(W), \tag{15}$$

with an obvious analogue holding for Whitney sums of vector bundles. If λ_t is applied to (13) one gets $\lambda_t(n\xi) = \lambda_t(\eta \oplus s\epsilon)$, or

$$(1 + \xi t)^n = \lambda_t(\eta)\lambda_t(s\epsilon) = \lambda_t(\eta)(1 + t)^s$$

where the 1 actually stands for the trivial line bundle. Introduce a new formal variable τ, and substitute the t above by

$$t = \frac{\tau}{1 - \tau} = \tau + \tau^2 + \tau^3 + \cdots$$

Then as power series,

$$\left(1 + \frac{\xi\tau}{1 - \tau}\right)^n = \lambda_{\frac{\tau}{1-\tau}}(\eta)\left(1 + \frac{\tau}{1 - \tau}\right)^s$$

or, after multiplying both sides with $(1 - \tau)^n$:

$$(1 + (\xi - \epsilon)\tau)^n = (1 - \tau)^{n-s}\lambda_{\frac{\tau}{1-\tau}}(\eta) \tag{16}$$

We remark first that the expression $\xi - \epsilon$, being a formal difference of two vector bundles over $\mathbb{R}P^{r-1}$, can be interpreted as an element in the Grothendieck ring $KO(\mathbb{R}P^{r-1})$, a standard practice in K-theory. In fact, since $\dim \xi = 1 = \dim \epsilon$, $x = \xi - \epsilon$ can be interpreted as an element in the reduced ring $\widetilde{KO}(\mathbb{R}P^{r-1})$. Secondly, since the vector bundle η has dimension $n - s$, $\lambda_t(\eta)$ is a polynomial in t of degree $\leq n - s$, and so the right hand side of (16) is a polynomial in τ with degree $\leq n - s$. Comparing coefficients we conclude that

$$\binom{n}{n - s + 1}x^{n-s+1}, \binom{n}{n - s + 2}x^{n-s+2}, \ldots$$

should be zero elements in $\widetilde{KO}(\mathbb{R}P^{r-1})$. This is Atiyah's necessary condition for Φ to exist. Notice that it is, in a sense, also a kind of height condition, as in the Stiefel-Hopf theorem.

Remark 2. If one is reluctant to use K-theory language, a condition like $\binom{n}{q}x^q = 0$ in $\widetilde{KO}(\mathbb{R}P^{r-1})$ could be re-interpreted in the following way: Expanding $x^q = (\xi - \epsilon) \otimes \cdots \otimes (\xi - \epsilon)$ binomially and moving negative terms to the right-hand side yields

$$\binom{n}{q}\left[\xi^q \oplus \binom{q}{2}\xi^{q-2} \oplus \cdots\right] =^s \binom{n}{q}\left[\binom{q}{1}\xi^{q-1} \oplus \binom{q}{3}\xi^{q-3} \oplus \cdots\right] \quad (17)$$

where ξ^q means $\xi \otimes \xi \cdots \otimes \xi$ (q times) and the symbol $=^s$ means that the vector bundles on both sides of (17) will become isomorphic after forming Whitney sum with ϵ for sufficiently many times.

Remark 3. However, the language of K-theory is more effective to help recast our condition into better form. First note that

$$\begin{aligned} x^2 &= (\xi - \epsilon) \otimes (\xi - \epsilon) \\ &= \xi \otimes \xi - \xi \otimes \epsilon - \epsilon \otimes \xi + \epsilon \otimes \epsilon \\ &= -2(\xi \otimes \epsilon - \epsilon) = -2(\xi - \epsilon) = -2x \end{aligned}$$

because $\xi \otimes \epsilon \cong \xi \cong \epsilon \otimes \xi$ and $\xi \otimes \xi \cong \epsilon \cong \epsilon \otimes \epsilon$. Inductively, then $x^q = (-1)^{q-1}2^{q-1}x$ and the condition simply says that the integers

$$2^{n-s}\binom{n}{n-s+1}, 2^{n-s+1}\binom{n}{n-s+2}, \ldots \quad (18)$$

are all divisible by the order of the element $x \in \widetilde{KO}(\mathbb{R}P^{r-1})$. Summarizing, one has:

Theorem 2. *If there exists a nonsingular bilinear may* $\Phi : \mathbb{R}^r \times \mathbb{R}^s \to \mathbb{R}^n$, *then necessarily the integers in (18) should all be divisible by* $2^{\phi(r-1)}$ *where*

$$\phi(r-1) = \#\{t : 1 \leq t \leq r-1 \ and \ t \equiv 0, 1, 2, \ or \ 4 (mod \ 8)\}$$

This is so simply because $\widetilde{KO}(\mathbb{R}P^{r-1})$ is well-known to be a cyclic group of order $2^{\phi(r-1)}$ [1].

Corollary 1. *If there exists a nonsingular bilinear map of type* (r, n, n), *then necessarily* n *is divisible by* $2^{\phi(r-1)}$

Illustration. Take $r = 11, s = 13$ and $n = 16$. Then $\phi(r-1) = 6$ and $2^{n-s}\binom{n}{n-s+1} = 2^3\binom{16}{4} = 2^5 \times 5 \times 7 \times 13$ is not divisible by 2^6. Therefore there cannot exist a nonsingular bilinear map of type (11, 13, 16). Accordingly any homogeneous bilinear system of 16 equations in $x_1, \ldots, x_{11}; y_1, \ldots, y_{13}$ admits a nontrivial solution. Note that this conclusion is not deducible from the Stiefel-Hopf theorem. In fact, various possible improvements to that theorem, such as Atiyah's here, are still being sought by topologists today. See §3.3 of Lecture 3.

Exercise 11. Why is the above corollary already optimal in a generic sense?

Exercise 12. The Stiefel-Hopf condition is symmetrical in r and s. What about the Atiyah's condition?

3. Lecture

3.1 *The Hopf construction and bilinear representation of homotopy classes of spheres*

To any nonsingular bilinear map Φ of type (r, s, n), there is a corresponding map

$$J_\phi : S^{r+s-1} \to \mathbb{R}^{n-1}\backslash\{0\}$$

defined explicitly by

$$J_\phi(x_1, \ldots, x_r; y_1, \ldots, y_s) = (|x|^2 - |y|^2, 2\Phi(x, y)) \tag{19}$$

where $|x|^2$ means the Euclidean norm square $x_1^2 + \cdots + x_r^2$. This is called the Hopf-construction (or the J construction) on Φ. It plays an important role in the homotopy theory of spheres. For example, if Φ is the multiplication of $\mathbb{R}, \mathbb{C}, \mathbb{H}$, or $\mathbb{O}$, J_ϕ is respectively,

$$\begin{aligned}
2\iota : S^1 &\to S^1 \subset \mathbb{R}^2\backslash\{0\} (\text{ Same as } z \mapsto z^2)\\
\eta : S^3 &\to S^2 \subset \mathbb{R}^3\backslash\{0\}\\
\nu : S^7 &\to S^4 \subset \mathbb{R}^5\backslash\{0\}\\
\sigma : S^{15} &\to S^8 \subset \mathbb{R}^9\backslash\{0\}
\end{aligned}$$

The last three maps, when suitably suspended, furnish generators for the stable 1,3,7-stem in the stable homotopy theory of spheres. We say that these generators are bilinearly representable, because their description can ultimately be traced back to certain nonsingular bilinear maps Φ.

As it turns out, many more stable homotopy classes of spheres are representable in this way, which is another reason why nonsingular bilinear maps are worth studying. We shall illustrate this through an example below.

For $k \leq n$, let $F_{n,k}$ be the space of all $k \times n$ real matrices of full rank k. Consider the map

$$\mathbb{R}^2\backslash\{0\} \to F_{3,2}$$

defined by

$$A(x_1, x_2) = \begin{bmatrix} x_1^3 & x_2 & 0 \\ x_2 & x_1 & x_2 \end{bmatrix}.$$

The map $g : \mathbb{R}^2 \times \mathbb{R}^2 \to \mathbb{R}^3$ given by

$$g((x_1, x_2); (y_1, y_2)) = (y_1, y_2)A(x_1, x_2)$$

is nonsingular and linear in y, but is not bilinear because not all the entries of $A(x_1, x_2)$ are linear forms in x_1 and x_2. Since $\mathbb{R}^4\backslash\{0\}$ has S^3 as a deformation retract, the Hopf construction $J_g : S^3 \to \mathbb{R}^4\backslash\{0\}$ represents a homotopy class $\alpha \in \pi_3(S^3)$, but α is not bilinearly represented by g as such. To claim that α is nevertheless a source for nonsingular bilinear maps, we perform, stably, a linearization procedure in the x variables as follows. Prolongate A to the map

$$A^+ : \mathbb{R}^2\backslash\{0\} \to F_{7,6}$$

defined by

$$A^+(x_1, x_2) = \begin{bmatrix} A(x_1, x_2) & 0 \\ 0 & I_4 \end{bmatrix}$$

where I_4 is the 4×4 identity matrix. The map

$$g^+ : \mathbb{R}^2 \times \mathbb{R}^6 \to \mathbb{R}^7$$

given by $g^+((x_1 x_2); (y_1, \ldots, y_6) = (y_1, \ldots, y_6)A^+(x_1, x_2)$ has Hopf construction

$$J_g^+ : S^7 \to \mathbb{R}^8\backslash\{0\}.$$

It represents a homotopy class $\alpha^+ \in \pi_7(S^7)$ which can be easily shown to be the 4-fold suspension of α. Thus α^+ and α define the same stable homotopy class of spheres.

For $0 \le t \le 1$, let $B_t : \mathbb{R}^2\backslash\{0\} \to F_{7,6}$ be defined by

$$B_t(x_1, x_2) = \begin{bmatrix} x_1^3 & x_2 & 0 & t(x_1^3 - 2x_1) & \sqrt{2}tx_1^2x_2 & 0 & 0 \\ x_2 & x_1 & x_2 & -tx_2 & 0 & 0 & 0 \\ tx_1 & 0 & 0 & x_1 & \sqrt{2}x_2 & 0 & 0 \\ 0 & 0 & 0 & -\sqrt{2}x_2 & x_1 & 0 & 0 \\ 0 & 0 & 0 & 0 & 0 & x_1 & -\sqrt{2}x_2 \\ 0 & 0 & 0 & 0 & 0 & \sqrt{2}x_2 & x_1 \end{bmatrix}$$

Note that this 6×7 matrix can be block-partitioned in the same way as $A^+(x_1, x_2)$ and that

(i) $B_t(x_1, x_2)$ has rank 6 for all $0 \le t \le 1$ and $(x_1, x_2) \in \mathbb{R}^2\backslash\{0\}$. This is not difficult to check after one notes that the third row is always perpendicular to the second row and to the bottom three rows.

(ii) For $t = 0$, $B_0(x_1, x_2)$ is in the same block partition format as $A^+(x_1, x_2)$. In fact the maps B_0 and A^+ are homotopic to each other.

(iii) On the other hand B_0 and B_1 are homotopic to each other, with $\{B_t\}_{0 \le t \le 1}$ providing the homotopy.

(iv) Let $L(x_1, x_2)$ be obtained from $B_1(x_1, x_2)$ by adding x_1^2 times the third row to the first row. This gives a map $L : \mathbb{R}^2\backslash\{0\} \to F_{7,6}$ homotopic to B_1. In symbols, we have now established

$$A^+ \sim B_0 \sim B_1 \sim L.$$

But the entries of $L(x_1, x_2)$ are, by inspection, linear forms in x_1, x_2, so the associated map $\Psi : \mathbb{R}^2 \times \mathbb{R}^6 \to \mathbb{R}^7$ given by

$$\Psi((x_1, x_2), (y_1, \ldots, y_6)) = [y_1, \ldots, y_6]L(x_1, x_2)$$

is nonsingular bilinear, while $J\Psi = \alpha^*$ holds. This shows that the original homotopy class α of spheres is (stably) representable, via a nonsingular bilinear Ψ which has just been constructed.

While α is a familiar homotopy class in the zero stem, there are many other esoteric homotopy classes α' of spheres which are likewise bilinearly representable by a corresponding Ψ' again through some linearization process. Such Ψ' remains as mysterious as α' is esoteric.

Exercise 13. Let $\eta : S^3 \to S^2$ be the Hopf construction on the multiplication map of $\mathbb{C}$. The composite

$$S^4 \xrightarrow{\Sigma\eta} S^3 \xrightarrow{\eta} S^2$$

where Σ means suspension, defines a nontrivial homotopy class of spheres η^2 in the stable 2-stem. Show that η^2 is bilinearly representable.

3.2 *The topology and geometry of truncated projective spaces*

The standard embedding

$$\mathbb{R}^1 \subset \mathbb{R}^2 \subset \cdots \subset \mathbb{R}^{m+1} \subset \cdots$$

leads to the inclusions

$$\{\text{point}\} = \mathbb{R}P^0 \subset \mathbb{R}P^1 \subset \cdots \subset \mathbb{R}P^m \subset \cdots$$

from which one gets the cellular structure for $\mathbb{R}P^m$:

$$\mathbb{R}P^m = \{\text{point}\} \cup e^1 \cup \cdots \cup e^m$$

where $e^i = \mathbb{R}P^i \backslash \mathbb{R}P^{i-1}$ is an i-dimensional cell. If $n < m$ and if we identify the subspace $\mathbb{R}P^{n-1}$ to one point, the identification space

$$\mathbb{R}P_n^m = \mathbb{R}P^m / \mathbb{R}P^{n-1} = \{\text{point}\} \cup e^n \cup \cdots \cup e^n$$

is called a truncated projective space. Included inside, of course, is its subspace

$$\mathbb{R}P^n / \mathbb{R}P^{n-1} = \{\text{point}\} \cup e^n.$$

This is homeomorphic to S^n. By a slightly incorrect terminology, this S^n is sometimes referred to as the bottom cell of the truncated projective space.

Exercise 14.

(i) Show that $\mathbb{R}P^1$ is homeomorphic to a circle and $\mathbb{R}P^3$ is homeomorphic to $SO(3)$, the space of all 3×3 special orthogonal matrices. (Hint: one approach is to make use of quaternions.)

(ii) Describe the subspace of $SO(3)$ which corresponds to the subspace $\mathbb{R}P^2$ of $\mathbb{R}P^3$ under the homeomorphism in part (i).

Exercise 15. For a vector bundle E over a compact base space, the *Thom space* $T(E)$ means the one-point compactification of the total space of E. Show that the Thom space of $n\xi_{r-1}$ over $\mathbb{R}P^{r-1}$ is homeomorphic to $\mathbb{R}P_n^{n+r-1}$.

Exercise 16. Recall that if (X, x_0), (Y, y_0) are topological spaces with base points x_0 and y_0, then the subspace of $X \times Y$:

$$\{(x, y) \in X \times Y : x = x_0, \text{ or } y = y_0\}$$

is called the wedge product (or one-point union) of X and Y, and is denoted by $X \vee Y$. Decompose the truncated projective space $\mathbb{R}P_2^3$ as a wedge product of two topological spaces.

It is now possible to use the Hurwitz-Radon family in §1.1 to prove:

Theorem 3. *For $r \geq 2$, the truncated projective space $\mathbb{R}P_n^{n+r-1}$ has the same homotopy type as $S^n \vee \mathbb{R}P_{n+1}^{n+r-1}$, provided that $r \leq \rho(n)$ where ρ is the Hurwitz-Radon function.*

Exercise 17. Give a detailed proof of this theorem based on the previous few exercises.

A much deeper result on the topology of truncated projective spaces is, however, obtained by J.F. Adams, in the converse theorem below.

Theorem 4 (Adams). *If $\mathbb{R}P_n^{n+r-1}$ has the same homotopy type as $S^n \vee \mathbb{R}P_{n+1}^{n+r-1}$, then necessarily $r \leq \rho(n)$.*

The proof uses K-theory and exterior power operations on vector bundles. See Adams [1]. These two theorems give a necessary and sufficient condition for a truncated projective space to split off its bottom cell as a wedge summand. Adams' proof established, in particular, the following:

Corollary 2. *If there exists a nonsingular bilinear map $\Phi : \mathbb{R}^r \times \mathbb{R}^n \to \mathbb{R}^n$ then necessarily $r \leq \rho(n)$.*

3.3 *Strengthening the Stiefel-Hopf condition: the case of equal height [6]*

The Stiefel-Hopf theorem leaves open the question whether a nonsingular bilinear map of type $(r, s, r \circ s)$ can exist. To address this, we extend the theorem to

Theorem 5. *Let n stand for $r \circ s$. If there exists a nonsingular bilinear map $\Phi : \mathbb{R}^r \times \mathbb{R}^s \to \mathbb{R}^n$, then necessarily $n \geq r + s - \rho(n)$ where ρ is the Hurwitz-Radon function.*

Proof. We first point out a numerical fact: for $n = r \circ s$, the binomial coefficient $\binom{n-1}{r-1}$ is always odd. We shall also assume, without any harm, that $r \leq s$. Define from Φ a map

$$F : S^{r-1} \times S^{s-1} \to S^{n-1}$$

given by

$$F(x, y) = \frac{1}{N}(\Phi_1(x, y), \ldots, \Phi_n(x, y),$$

where $x = (x_1, \ldots, x_r)$, $y = (y_1, \ldots, y_s)$ are unit vectors and N is the norm of the nonzero vector $(\Phi_1(x, y), \ldots, \Phi_n(x, y)) \in \mathbb{R}^n$. Clearly $F(-x, y) = F(x, -y) = -F(x, y)$, so that F factors through the quotient space $S^{r-1} \times_T S^{s-1}$ obtained by identifying each point (x, y) in $S^{r-1} \times S^{s-1}$ with its "antipodal point" $T(x, y) = (-x, -y)$. One may think of $S^{r-1} \times_T S^{s-1}$ as a kind of "projective torus" and note that it has on its own self a fixed-point-free involution given by

$$\tau([x, y]) = [-x, y] = [x, -y].$$

When corresponding points under τ are further identified, the quotient of $S^{r-1} \times_T S^{s-1}$ clearly is the Cartesian product of two real projective spaces. The entire situation can be summarized by the diagram below, where vertical arrows denote identification maps, which are also double covering maps, and where $\tilde{F}$ and PF are maps induced by F.

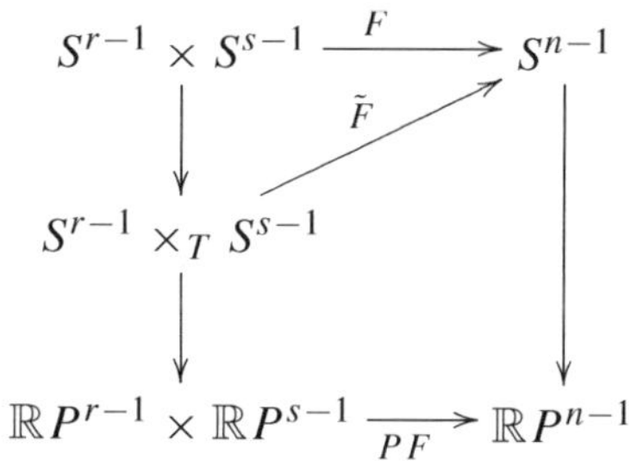

Let u, v, w be the generators of the 1-dimensional cohomology groups with mod2 coefficients of $\mathbb{R}P^{r-1}, \mathbb{R}P^{s-1}, \mathbb{R}P^{n-1}$ respectively. Consider the ring homomorphism induced by PF :

$$PF^* : H^*(\mathbb{R}P^{n-1}; \mathbb{Z}_2) \to H^*(\mathbb{R}P^{r-1} \times \mathbb{R}P^{s-1}; \mathbb{Z}_2). \tag{20}$$

By restricting PF to $\mathbb{R}P^0 \times \mathbb{R}P^{s-1}$ and $\mathbb{R}P^{r-1} \times \mathbb{R}P^0$ Hopf argued that $(PF)^*(w) = u \otimes 1 + 1 \otimes v$. If n had been strictly less than $r \circ s$, then this would be a contradiction because $(PF)^*$ cannot increase height. (This was a main step in Hopf's proof of his theorem). But in the present situation n is equal to $r \circ s$, and there is no contradiction because w and $u \otimes 1 + 1 \otimes v$ are of equal height $r \circ s$. In such a case of "equal height", however, various strong versions of Borsuk-Ulam theorem will come into play. For our purpose a most suitable version is

Theorem 6 (of Borsuk-Ulam type). *Let $\tilde{W}$ be a compact connected cell complex with a fixed-point-free involution simply denoted by $\tilde{w} \leftrightarrow -\tilde{w}$ and let W be the quotient space obtained by identifying antipodal points. Suppose $\tilde{F} : \tilde{W} \to S^{n-1}$ is an antipodal map so that under its projectivisation $P\tilde{F} : W \to \mathbb{R}P^{n-1}$ one has the condition that:*

> *the generator $w \in H^1(\mathbb{R}P^{n-1}; \mathbb{Z}_2)$ and the element $(PF)^*(w) \in H^1(W; \mathbb{Z}_2)$ are of equal height.* (21)

Then $\tilde{F}^* : H^{n-1}(S^{n-1}; \mathbb{Z}_2) \to H^{n-1}(\tilde{W}; \mathbb{Z}_2)$ must be a nonzero homomorphism.

A good proof of this theorem is to use the "transfer" map associated to double coverings. One can consult Bredon's book "*Topology and geometry*" chapter IV, §20 for full details. Applying this to $\tilde{F}$ in the previous diagram, we now conclude that

$$\tilde{F}^* : H^{n-1}(S^{n-1}; \mathbb{Z}_2) \to H^{n-1}(S^{r-1} \times_T S^{s-1}; \mathbb{Z}_2)$$

must be nontrivial, or that $H^{n-1}(S^{r-1} \times_T S^{s-1}; \mathbb{Z}_2)$ must have a nonzero cospherical cohomology class in dimension $n-1$.

Let us discuss the geometry of $\tilde{Y} = S^{r-1} \times_T S^{s-1}$. If again ξ denotes the canonical Hopf line bundle over $\mathbb{R}P^{r-1}$, then the vector bundle $s\xi$ has total space $S^{r-1} \times_T \mathbb{R}^s$, and $\tilde{Y}$ is its corresponding unit sphere bundle. Since we are assuming $r \leq s$, $s\xi$ admits a unit section $\sigma : \mathbb{R}P^{r-1} \to \tilde{Y}$. It thus splits up into a Whitney sum $\eta \oplus \epsilon$ where ϵ denotes the trivial line bundle whose fiber over $[x]$ is spanned by $\sigma[x]$, and η is a vector bundle of dimension $s-1$. By elementary point set topology, it is easy to verify, using stereographic projection, that the Thom space $T(\eta)$ of η is homeomorphic to $\tilde{Y}/\sigma(\mathbb{R}P^{r-1})$, the space obtained from $\tilde{Y}$ by identifying the image of the entire section σ to a point. By standard properties of Thom complexes, we now have homeomorphisms

$$T(s\xi) \approx T(\eta \oplus \epsilon) \approx \Sigma(T(\eta)) \approx \Sigma(\tilde{Y}/\sigma(\mathbb{R}P^{r-1})),$$

exhibiting $\tilde{Y}/\sigma(\mathbb{R}P^{r-1})$ as a 'desuspension' of the Thom space $T(s\xi)$ which by Exercise 15 in §3.1, is homeomorphic to $\mathbb{R}P_s^{r+s-1}$.

Returning to the proof of theorem 5, in the above diagram, we note that the map $\tilde{F}$ is homotopically trivial when restricted to $\sigma(\mathbb{R}P^{r-1})$. This is simply for reason of dimension, because $r \leq s < n$. (Theorem 5 in the case $s = n$ is already dealt with by Atiyah's result, see the corollary in §2.3). Thus $\tilde{F}$ factors up to homotopy through a map $\hat{F}$ as in the following diagram.

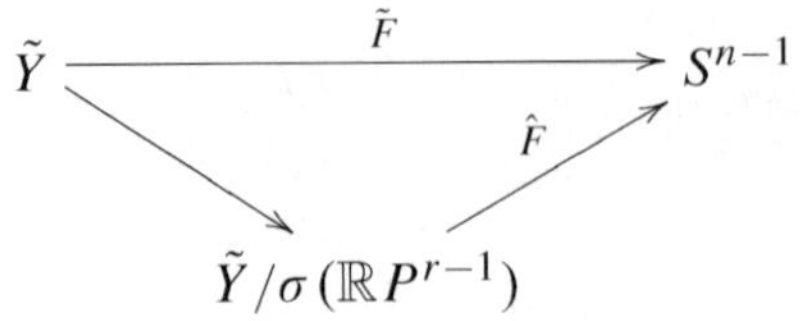

Suspending this diagram once and using the homeomorphisms above, we get a homotopy commutative diagram

$$\begin{array}{ccc} \Sigma\tilde{Y} & \xrightarrow{\quad\Sigma\tilde{F}\quad} & S^n \\ & \searrow \qquad \nearrow_{\Sigma\hat{F}} & \\ & \mathbb{R}P_s^{r+s-1} & \end{array}$$

in which $(\Sigma\tilde{F})^*$ is nontrivial in n-dimensional cohomology. The same must be true, by commutativity, for $(\Sigma\hat{F})^*$ as well. We can now use cellularity to identify the $(n-1)$-skeleton of $\mathbb{R}P_s^{r+s-1}$ to a point, obtaining from $(\Sigma\hat{F})$ a quotient map $\bar{F}$: $\mathbb{R}P_n^{r+s-1} \to S^n$ which exhibits the cosphericity of the nonzero cohomology class in $H^n(\mathbb{R}P_n^{r+s-1}; \mathbb{Z}_2)$. In other words, $\bar{F}$ shows, in Bott's terminology, that the bottom cell of $\mathbb{R}P_n^{r+s-1}$ is co-reducible. Using standard argument, $\bar{F}$, serving as a "retraction map", ensures that $\mathbb{R}P_n^{r+s-1}$ has the same homotopy type as $S^n \vee \mathbb{R}P_{n+1}^{r+s-1}$. By Adams' theorem 4 in §3.1, one obtains $r+s-n \leq \rho(n)$, or $r \circ s \geq r+s-\rho(r \circ s)$, as is to be proved. Q.E.D.

Illustration. For $r = 20$ and $s = 23$, $r \circ s$ is equal to 32 and $\rho(r \circ s) = 10$. Any bilinear map of type (20, 23, 32) must therefore be singular.

Remark 4. Theorem 5 provides a necessary condition on r and s for Φ to exist. That condition is not yet sufficient though. For example, if $r = 11$, $s = 13$, then $r \circ s = 16$ and $\binom{r \circ s - 1}{r-1}$ is odd, but Atiyah's result [See illustration in §2.3] says a nonsingular bilinear map of type (11, 13, 16) cannot exist. In fact, up to this day, it is still a challenging open question to determine all r, s such that there can exist a nonsingular bilinear map of type $(r, s, r \circ s)$!

Editor's note: For contents of the fourth lecture see the article *'Orthogonal Multiplications and minimal submanifolds of spheres'* by the author in this volume.

References

[1] J. F. Adams, Vector fields on Spheres, *Ann. of Math. (2)*, **75** (1962) 603–632.

[2] M. F. Atiyah, Immersion and embedding of manifolds, *Topology*, **1** (1962) 125–132.

[3] F. Behrend, Über Systeme reeler algebraisher Gleichungen, *Composito Math.*, **7** (1939) 1–19.

[4] H. Hopf, Ein topologicher Beitrag zur reelen Algebra, *Commentarri Math. Helv.*, **13** (1941) 219–239.

[5] K. Y. Lam, On bilinear and skew-linear maps that are nonsingular, *Quart. J. Math. Oxford (2)*, **19** (1968) 281–288.

[6] K. Y. Lam, Borsuk-Ulam type theorems and systems of bilinear equations, Geometry from the Pacific Rim, Walter de Gruyter, (1997) 183–194.

[7] N. E. Steenrod, The topology of fibre bundles, Princeton University Press (1951).

Perspectives in Geometry and Topology No. 11, 2010, pp. 49–65.

On Algebraic Deformation Theory

Goutam Mukherjee

Stat-Math Unit, Indian Statistical Institute, Kolkata
e-mail: goutam@isical.ac.in

Abstract. The aim of this note is to give a brief account of algebraic deformation theory. We begin with the classical work of Gerstenhaber on formal one parameter deformation theory of associative algebras and explain its modern approach by considering deformations of Leibniz algebras over a commutative local algebra. At the end, we give a construction of versal deformation in the context of Leibniz algebras.

Keywords. Deformation theory, Algebra, infinitesimal deformation, Hochschild Cohomology, Leibniz algebra, Versal deformation.

1. Introduction

Deformation theory dates back at least to Riemann's 1857 memoir on abelian functions in which he studied manifolds of complex dimension one and calculated the number of parameters (called moduli) upon which a deformation depends. The modern theory of deformations of structures on manifolds was developed extensively by Frolicher-Kodaira-Nijenhuis-Nirenberg-Spencer [14], [15], [13], [26], [27]. The study of deformations of algebraic structures was initiated by Gerstenhaber in a series of papers [7], [8], [9], [10], [11]. He introduced formal one parameter deformation theory for associative algebras and remarked that his methods would extend to any equationally defined algebraic structure.

Intuitively a deformation of a mathematical object is a family of the same kind of objects depending on some parameter(s). Thus for example a family of real three dimensional algebras defined by $A_\lambda = \mathbb{R}[X]/(x^3 - \lambda x)$, $\lambda \in \mathbb{R}$. Of course the parameter, on which the family under consideration depends, need not be scalar but can be a vector which varies over a subset of a vector space or over more general objects such as algebraic schemes. By viewing the object associated to the parameter value λ as 'lying over λ' one obtains a 'fibred object'

$$X \longrightarrow B$$

where the fibre over $\lambda \in B$ is the object from the family labeled by λ. For example in the case of the family of three dimensional algebras mentioned above the fibre over λ is the commutative algebra $\mathbb{R}[X]/(x^3 - \lambda x)$. Thus for the geometric type categories such as topological spaces, schemes, manifolds with singularities, etc., a deformation is simply a (surjective) morphism with special stress on aspects and questions which involve how the fibres over $\lambda \in B$, vary with λ.

In more precise terms, any deformation theory should have the following features:

(i) A class of objects within which deformation takes place and a cohomology theory associated to those objects which control the deformation in the sense that infinitesimal deformation of a given object can be identified with the elements of a suitable cohomology group.
(ii) A theory of obstruction to the "integration" of an infinitesimal deformation.
(iii) Deformation of automorphisms of the deformed object and a notion of rigid objects.

For example, in Gerstenhaber theory, objects are associative algebras and the natural candidate for the cohomology is the Hochschild cohomology.

The main idea of Gerstenhaber is that given an associative algebra A, one deforms its multiplication while keeping the associativity property and its underlying additive structure. The infinitesimal of a deformation can then be identified with a 2-cocycle in the Hochschild cohomology of A with coefficients in itself. As a result, vanishing of this cohomology group implies that A is rigid, in the sense that every deformation is equivalent to the original multiplication. The next question is to ask whether a given infinitesimal can be extended to a full-blown deformation. Suppose a given infinitesimal has been extended to a finite order deformation (say, to a deformation of order k). In order to extend it to a $(k + 1)$-th order deformation, one encounters an obstruction which is identified as a 3-cocycle, the vanishing of the corresponding cohomology class is a necessary and sufficient condition for the extension to be possible.

The methods developed by Gerstenhaber have been applied successfully to other categories of algebras. See for examples, [22], [23], [24] for Lie algebras, and [12] for the deformation theory of Hopf algebras which relate to quantum groups. See also [25,6] for related works. For some more recent related works, see, [21], [19], [29–31].

In recent time, when deformations of various algebraic objects are being studied, it is felt that it is desirable to have a theory of deformations over more general base, not merely, over formal power series ring. Roughly speaking, "base" of a deformation is the space of parameters upon which the deformation takes place. For instance, in the case of formal one parameter deformation, the base is a formal power series ring $\mathbb{K}[[t]]$, where $\mathbb{K}$ is a field of characteristic zero (see Section 4 for details). The main problem in deformation theory is to describe all non-equivalent deformations of a given object. The right approach to address this question is to develop a theory of deformation over a local algebra base, or more generally, over a complete local algebra base and to construct the so called 'versal deformation' – a deformation which induces all non-equivalent ones, in the sense that, any deformation is equivalent to a deformation obtained as a push out of versal deformation (Definition 5.6). The existence of such a "versal" deformation for algebraic categories follows from the work of Schlessinger [28]. For Lie algebras it was worked out in [3,5], and for Leibniz algebras it has been constructed recently in [4]. It turns out that (under some minor cohomology restrictions) there exists a versal element. It may be noted that a versal deformation, even if it exists, need not be unique. Sometimes in the literature, it is called universal in case it exists uniquely. An infinitesimal deformation is basically a first order deformation. Although versal deformation need not be unique, there exists an universal element in the class of all infinitesimal deformations over finite dimensional base (see Section 5 for details). Versal deformations have certain advantage over the classical notion of one parameter family of formal deformations of algebras as introduced by Gerstenhaber, as in the classical theory one cannot recover all non-equivalent deformations of a given object. In contrast, versal deformation induces all other deformations.

We first briefly describe the classical case of formal one parameter deformation theory of associative algebras as developed by Gerstenhaber. In the subsequent sections we describe the notion of deformation over a local algebra base for Leibniz algebras and give a sketch of the construction of a versal deformation for Leibniz algebras.

2. Formal one parameter deformations of associative algebras

Let A be an associative algebra over a field $\mathbb{K}$ of characteristic zero. Let us recall the definition of Hochschild cohomology of A with coefficients in A, regarded as an A-bimodule.

Definition 2.1. *Let A be a $\mathbb{K}$-algebra. A bimodule M over A (or, briefly, an A-bimodule) is a $\mathbb{K}$-module equipped with two actions (left and right) of A*

$$A \times M \longrightarrow M \text{ and } M \times A \longrightarrow M$$

such that $(am)a' = a(ma')$ for $a, a' \in A$ and $m \in M$.

The actions of A and $\mathbb{K}$ on M are compatible, that is

$$(\lambda a)m = \lambda(am) = a(\lambda m)$$

and

$$m(\lambda a) = (\lambda m)a = \lambda(ma),$$

for $\lambda \in \mathbb{K}$, $a \in A$ and $m \in M$. When A has the identity 1 we always assume that $1m = m1 = m$ for $m \in M$.

Given a bimodule M over A, the Hochschild complex of A with coefficients in M is defined as follows.

Set $C_n(A, M) = M \otimes A^{\otimes n}; n \geq 0$ where $A^{\otimes n} = A \otimes \cdots \otimes A$ (n-copies). Let

$$\partial : C_n(A, M) \longrightarrow C_{n-1}(A, M)$$

be the $\mathbb{K}$-linear map given by

$$\partial(m, a_1, a_2, \ldots, a_n) = (ma_1, a_2, \ldots, a_n) + \sum_{i=1}^{n-1}(-1)^i(m, a_1, \ldots, a_i a_{i+1}, \ldots, a_n) + (-1)^n(a_n m, a_1, \ldots, a_{n-1}). \quad (1)$$

Then $\partial^2 = 0$ and the complex $(C_*(A, M), \partial)$ is called the Hochschild complex of A with coefficients in the A-bimodule M. When $M = A$, where the actions are given by algebra operation in A we denote the complex $C_*(A, A)$ by $C_*(A)$. For Gerstenhaber's theory we need Hochschild cohomology of A with coefficients in A, regarded as an A-bimodule.

The cochain complex is defined as follows. Let $C^n(A) = Hom_{\mathbb{K}}(C_n(A), A)$ and the coboundary $\delta^n : C^n(A) \longrightarrow C^{n+1}(A)$ is given by

$$\begin{aligned}
&\delta^n(f)(a_1, \ldots, a_{n+1}) \\
&\quad = a_1 f(a_2, \ldots, a_{n+1}) + \sum_{0<i<n+1} (-1)^i f(a_1, \ldots, a_i a_{i+1}, \ldots, a_{n+1}) \\
&\quad + (-1)^{n+1} f(a_1, \ldots, a_n) a_{n+1}.
\end{aligned}$$

Then $\delta^{n+1} \circ \delta^n = 0$ and we have the Hochschild cochain complex $(C^*(A), \delta^*)$, the corresponding cohomology is denoted by $H^*(A, A)$.

For deformation theory, we assume that the field $\mathbb{K}$ is of characteristic zero. Let $K = \mathbb{K}[[t]]$ be the formal power series ring with coefficients in $\mathbb{K}$.

Definition 2.2. *A formal one parameter deformation of A is a K-bilinear multiplication law $m_t : A[[t]] \times A[[t]] \longrightarrow A[[t]]$ on the space $A[[t]]$ of formal power series in a variable t with coefficients in A, satisfying the following properties:*

$$m_t(a, b) = m_0(a, b) + m_1(a, b)t + m_2(a, b)t^2 + \cdots \text{ for } a, b \in A,$$

where $m_i : A \times A \longrightarrow A$ are $\mathbb{K}$-bilinear and $m_0(a, b) = a \cdot b$ is the original multiplication on A, and m_t is associative.

Thus a formal deformation of A is an associative algebra structure m_t on $A[[t]]$ over K such that $A_0 = A[[t]] \otimes_K \mathbb{K}$ is isomorphic to A, where $\mathbb{K}$ is considered as a K-module via the augmentation map $K \longrightarrow \mathbb{K}$.

Note that the condition that m_t be associative is equivalent to the equation

$$m_t(m_t(a, b), c) = m_t(a, m_t(b, c)), \quad \text{for } a, b, c \in A.$$

Writing down the above equation explicitly, we get

$$\sum_{\substack{\lambda+\mu=\nu \\ \lambda,\mu\geq 0}} m_\lambda(m_\mu(a, b), c) - m_\lambda(a, m_\mu(b, c)) = 0, \ \nu \geq 0. \tag{2}$$

For $\nu = 0$, this is just the associativity of the original multiplication. For $\nu = 1$, the above condition implies

$$am_1(b, c) - m_1(ab, c) + m_1(a, bc) - m_1(a, b)c = 0.$$

In terms of Hochschild theory this simply means that m_1 is a 2-cocycle, that is, $m_1 \in Z^2(A, A)$. The 2-cochain m_1 in the above definition of a formal deformation of an associative algebra A is called the infinitesimal of the deformation and is always a 2-cocycle.

Thus we have

Proposition 2.3. *The infinitesimal of a one parameter family of formal deformation of an associative algebra is a Hochschild 2-cocycle.*

Remark 2.4. The base of a formal one parameter deformation as defined above is K. However, as remarked before in the introduction, deformation makes sense over more general base. For instance, if $(A[[t]], m_t)$ is a formal one parameter deformation of A, then m_t modulo t^2, where m_t is the formal multiplication as given in the above definition, is a deformation of A over the base $K/(t^2)$ and is called an infinitesimal deformation of A. More generally, a one parameter deformation of order n, is a deformation with base $K/(t^{n+1})$, given by m_t modulo t^{n+1}. In this case, the associativity condition means that the 2-cochains m_ν, satisfy (2) for $0 \leq \nu \leq n$.

Next comes the question of extending an infinitesimal deformation to a full-blown deformation. If we start with an arbitrary 2-cocycle $m_1 \in Z^2(A, A)$, it need not be an 'infinitesimal' of a deformation. If it be so, then we say that m_1 is integrable. The integrability of m_1 implies an infinite sequence of relations which may be interpreted as the vanishing of the 'obstructions' to the integration of m_1.

For suppose we have a deformation of A of order $n \geq 1$, g) for $0 \leq \nu \leq n$. In order that the given n^{th} order deformation extends to an $(n+1)^{th}$ order deformation m_t modulo t^{n+2}, over $K/(t^{n+2})$, the multiplication m_t modulo t^{n+2} must be associative, equivalently, (2) should hold for $0 \leq \nu \leq n+1$. Using coboundary formula, the equation for $\nu = n+1$ can be rewritten as

$$\delta m_{n+1}(a, b, c) = \sum_{\substack{\lambda+\mu=n \\ \lambda,\mu>0}} m_\lambda(m_\mu(a, b), c) - m_\lambda(a, m_\mu(b, c)). \tag{3}$$

In view of this observation, given an nth order deformation m_t modulo t^{n+1} over $K/(t^{n+1})$, we define a function G by

$$G(a, b, c) = \sum_{\substack{\lambda+\mu=n \\ \lambda,\mu>0}} m_\lambda(m_\mu(a, b), c) - m_\lambda(a, m_\mu(b, c)), \ a, b, c \in A.$$

Then G is a 3-cochain and is called the $(n+1)^{th}$ obstruction cochain in extending a deformation of order n to a deformation of order $n+1$. One of the main observations in deformation theory is the following result.

Theorem 2.5. *The $(n+1)^{th}$ obstruction cochain G is a 3-cocycle and the given n^{th} order deformation extends to a deformation of order $(n+1)$ if and only if the cohomology class of G vanishes.*

We immediately obtain

Corollary 2.6. *If $H^3(A, A) = 0$ then any Hochschild 1-cocyle is integrable.*

A one parameter family of deformations of an associative algebra A defined by a multiplication m_t is trivial if there exists a K-linear automorphism Φ_t of $A[[t]]$ given by

$$\Phi_t(a) = a + t\Phi_1(a) + t^2\Phi_2(a) + \cdots, \ a \in A$$

where each $\Phi_i : A \longrightarrow A$ are linear maps such that

$$m_t(a,b) = {\Phi_t}^{-1}(\Phi_t(a) \cdot \Phi_t(b)).$$

In that case, the deformed algebra is isomorphic to the algebra on $A[[t]]$ induced from that of A and by K-linearity.

There is an obvious notion of equivalence of deformations and an associative algebra A is called rigid if every deformation is trivial. The following is a consequence.

Theorem 2.7. *If $H^2(A, A) = 0$, then A is rigid.*

For example, it is well known that the tensor algebra $A = T(V)$ on any vector space V and the universal enveloping algebra $U(\mathcal{G})$ of a semi simple Lie algebra $\mathcal{G}$ are rigid.

3. Leibniz algebra and its cohomology

Definition 3.1. *A Leibniz algebra is a $\mathbb{K}$-vector space L, where $\mathbb{K}$ is any field, equipped with a bracket operation which is $\mathbb{K}$-bilinear and satisfies the Leibniz identity:*

$$[x,[y,z]] = [[x,y],z] - [[x,z],y], \quad x,\ y,\ z \in L.$$

Remark 3.2. The notion of Leibniz algebra makes sense when $\mathbb{K}$ is replaced by a ring R and L is an R-module.

The notion of Leibniz algebra was introduced by J. L. Loday [16] as a non-commutative (to be more precise, a non anti-symmetric) analogue of Lie algebra. Loday has investigated Leibniz algebras in connection with cyclic homology and Hochschild homology of matrix algebras. The details are available in [17,18]. Any Lie algebra, is automatically a Leibniz algebra, as in the presence of antisymmetry, the Leibniz identity reduces to the Jacobi identity. Here are some more examples:

Example 3.3. Let (L, d) be a differential Lie algebra, then $[x, y]_d := [x, dy]$, called the derived bracket, is a Leibniz bracket on L.

Example 3.4. On $\bar{T}(V) = V \oplus V^{\otimes 2} \oplus \cdots \oplus V^{\otimes n} \oplus \cdots$ there is a unique bracket which makes it into a Leibniz algebra and verifies

$$x_1x_2\ldots x_n = [\cdots[[x_1,x_2],x_3],\ldots,x_n],$$

where $x_1x_2\ldots x_n$ denotes the homogeneous element $x_1 \otimes x_2 \otimes \cdots \otimes x_n \in V^{\otimes n}$. Thus for instance

$$x_1x_2 = [x_1,x_2], \quad x_1x_2x_3 = [x_1x_2,x_3] = [[x_1,x_2],x_3]$$

and so on.

Example 3.5. Let A be any associative $\mathbb{K}$-algebra equipped with a $\mathbb{K}$-linear map $D : A \longrightarrow A$ satisfying the condition

$$D(a(Db)) = DaDb = D((Da)b), \quad a, b \in A.$$

Then, $[x, y] := x(Dy) - (Dy)x$ is a Leibniz bracket on A. Some examples of D satisfying the above identity are as follows:

1. D is an algebra map satisfying $D^2 = D$,
2. A is a superalgebra (i.e, A has a $\mathbb{Z}/2$-grading $A = A_+ \oplus A_-$ so that any $x \in A$ can be written uniquely as $x = x_+ + x_-$) and $D(x) = x_+$,
3. D is a square-zero derivation, i.e,

$$D(ab) = (Da)b + a(Db) \text{ and } D^2 a = 0.$$

Example 3.6. Some low dimensional examples of Leibniz algebras and their classifications are available now due to the work of Albeverio, Omirov and Rakhimov [1]. For instance, here is a 3-dimensional example. Let L be the space spanned by $\{e_1, e_2, e_3\}$ over $\mathbb{K}$. Define a bilinear map $[\ ,\ ] : L \times L \longrightarrow L$ by $[e_1, e_3] = e_2$ and $[e_3, e_3] = e_1$, and all other products of basis elements as 0. Then $(L, [\ ,\ ])$ is a Leibniz algebra over $\mathbb{K}$.

Example 3.7. Let A be any commutative, associative algebra over $\mathbb{K}$ and L a Leibniz algebra. Then $A \otimes \mathbb{K}$ is a Leibniz algebra over A.

To any Leibniz algebra L is associated a chain complex

$$CL_*(L) : \cdots L^{\otimes n} \stackrel{d}{\to} L^{\otimes n-1} \stackrel{d}{\to} \cdots \stackrel{d}{\to} L^{\otimes 2} \stackrel{d}{\to} L$$

where

$$d(x_1, \ldots, x_n) = \sum_{1 \le i < j \le n} (-1)^j (x_1, \ldots, x_{i-1}, [x_i, x_j], x_{i+1}, \ldots, \hat{x}_j, \ldots, x_n).$$

The homology groups of this complex are denoted by $HL_n(L)$, for $n \ge 1$.

Define $CL^n(L; L) := \mathrm{Hom}_{\mathbb{K}}(L^{\otimes n}, L)$, $n \ge 0$. Let

$$d^n : CL^n(L; L) \longrightarrow CL^{n+1}(L; L)$$

be a $\mathbb{K}$-homomorphism defined by

$$\begin{aligned} d^n f(x_1, \ldots, x_{n+1}) \\ := [x_1, f(x_2, \ldots, x_{n+1})] + \sum_{2}^{n+1} (-1)^i [f(x_1, \ldots, \hat{x}_i, \ldots, x_{n+1}), x_i] \\ + \sum_{1 \le i < j \le n} (-1)^{j+1} f(x_1, \ldots, x_{i-1}, [x_i, x_j], x_{i+1}, \ldots, \hat{x}_j, \ldots, x_n). \end{aligned}$$

Then $(CL^*(L; L), d)$ is a cochain complex, whose cohomology is called the cohomology of the Leibniz algebra L with coefficients in itself. The nth cohomology is denoted by $HL^n(L; L)$.

4. Deformations

Let L be a Leibniz algebra over a field $\mathbb{K}$ of characteristic zero and let A be a commutative algebra with identity over $\mathbb{K}$ with a fixed augmentation $\varepsilon : A \to \mathbb{K}$, which is an algebra homomorphism with $\varepsilon(1) = 1$. Let Ker $\varepsilon = \mathfrak{M}$. We assume that dim $(\mathfrak{M}^k / \mathfrak{M}^{k+1}) < \infty$ for all k. We abbreviate $\otimes_{\mathbb{K}}$ to $\otimes$.

Definition 4.1. *A deformation λ of L with base $(A, \mathfrak{M})$, is a A-Leibniz algebra structure on the tensor product $A \otimes L$ with the bracket $[\ ,\ ]_\lambda$, such that*

$$\varepsilon \otimes id : A \otimes L \to \mathbb{K} \otimes L = L$$

is a A-Leibniz algebra homomorphism, where the A-Leibniz algebra structure on $\mathbb{K} \otimes L$ is given via ε. If the algebra A is local such that the unique maximal ideal $\mathfrak{M}$ has the property, $\mathfrak{M}^2 = 0$, then any deformation with base A is called an infinitesimal deformation.

Remark 4.2. Observe that by A-linearity, to describe a deformation, it is enough to fix the elements $[1 \otimes x, 1 \otimes y]_\lambda$ for $x,\ y \in L$. Moreover,

$$\varepsilon \otimes id([1 \otimes x, 1 \otimes y]_\lambda) = [x, y] = \varepsilon \otimes id(1 \otimes [x, y]).$$

This implies, $[1 \otimes x, 1 \otimes y]_\lambda - 1 \otimes [x, y] \in$ Ker $(\varepsilon \otimes id)$. Hence we can write,

$$[1 \otimes x, 1 \otimes y]_\lambda = 1 \otimes [x, y] + \sum_i m_i \otimes x_i,$$

where $m_i \in \mathfrak{M}$ and $x_i \in L$ and $\sum_i m_i \otimes x_i$ is a finite sum.

Definition 4.3. *Let A be a complete local algebra, i.e, $A = \varprojlim(A/\mathfrak{M}^n)$ where $\mathfrak{M}$ is the maximal ideal in A. A formal deformation l of L with base A is a A-Leibniz algebra structure on the completed tensor product $A \hat{\otimes} L = \varprojlim((A/\mathfrak{M}^n) \otimes L)$ obtained as the projective limit of deformations $l_{A/\mathfrak{M}^n}$ with base $A/\mathfrak{M}^n$, such that $\varepsilon \otimes id : A \hat{\otimes} L \to L$, is an A-Leibniz algebra homomorphism.*

Example 4.4. If $A = \mathbb{K}[[t]]$, then a formal deformation of L with base A is the same as a formal 1-parameter deformation of L.

Definition 4.5. *Suppose λ_1 and λ_2 are two deformations of a Leibniz algebra L with base A. Then they are said to be equivalent if $\exists$ a Leibniz algebra isomorphism*

$$\phi : (A \otimes L,\ [\ ,\ ]_{\lambda_1}) \to (A \otimes L,\ [\ ,\ ]_{\lambda_2})$$

such that $(\varepsilon \otimes id) \circ \phi = \varepsilon \otimes id$.

We shall now introduce an important notion, called push out of a deformation.

Definition 4.6. *Let A' be another commutative algebra with identity and a fixed augmentation $\varepsilon' : A' \to K$. Suppose $\phi : A \to A'$ is an algebra homomorphism with $\phi(1) = 1$ and $\varepsilon' \circ \phi = \varepsilon$. Let $ker\ \varepsilon' = \mathfrak{M}'$. Suppose λ is a deformation of L with base $(A, \mathfrak{M})$, then the push-out $\phi_*\lambda$ is the deformation of L with base $(A', \mathfrak{M}')$ defined as follows. We regard A' as an A-module via ϕ, that is, $a' \cdot a = a' \cdot \phi(a)$. Then*

$$A' \otimes L = (A' \otimes_A A) \otimes L = A' \otimes_A (A \otimes L).$$

With the above identification, the deformation $\phi_*\lambda$ on $A' \otimes L$ is given by the bracket

$$[a_1{}' \otimes_A (a_1 \otimes l_1), a_2' \otimes_A (a_2 \otimes l_2)] = a_1' \cdot a_2' \otimes_A [a_1 \otimes l_1, a_2 \otimes l_2]_\lambda$$

where $a_1', a_2' \in A'$; $a_1, a_2 \in A$ and $l_1, l_2 \in L$.

It is easy to see that, if l is given as described in Remark 4.2, then

$$[1 \otimes x, 1 \otimes y]_{\phi_*\lambda} = 1 \otimes [x, y] + \sum_i \phi(m_i) \otimes x_i.$$

5. Universal infinitesimal deformation

Let L be a Leibniz algebra which satisfies the condition dim $(HL^2(L; L)) < \infty$. This is true, for example, if L is finite dimensional. Let us denote $HL^2(L; L)$ by H. Consider the algebra $A = \mathbb{K} \oplus HL^2(L; L)'$ by setting

$$(k_1, h_1).(k_2, h_2) = (k_1k_2, k_1h_2 + k_2h_1)$$

where $HL^2(L; L)'$ is the dual of $HL^2(L; L)$. Observe that second summand is an ideal of A with zero multiplication.

We are particularly interested in a special deformation μ_L of L with base $A = \mathbb{K} \oplus HL^2(L; L)'$. We fix a homomorphism

$$\mu : H \longrightarrow C^2(L; L) = Hom(L^{\otimes 2}; L)$$

which takes each cohomology class into a cocycle representing it. (This can be done easily first by assigning μ on a basis of H and then extending it linearly.) Observe that there is an isomorphism $H' \otimes L \cong Hom(H; L)$. Now define a Leibniz algebra structure on $A \otimes L$ as follows: we write

$$A \otimes L = (\mathbb{K} \oplus H') \otimes L = \mathbb{K} \otimes L \oplus H' \otimes L = L \oplus Hom(H; L).$$

Using the above identification, we define a bracket on $A \otimes L$ by

$$[(l_1, \phi_1), (l_2, \phi_2)] = ([l_1, l_2], \psi)$$

for $(l_1, \phi_1), (l_2, \phi_2) \in L \oplus Hom(H; L)$ where $\psi : H \longrightarrow L$ is defined by

$$\psi(\alpha) = \mu(\alpha)(l_1, l_2) + [\phi_1(\alpha), l_2] + [l_1, \phi_2(\alpha)] \quad \forall \alpha \in H.$$

It is routine to check that:

Proposition 5.1. *The vector space $A \otimes L$ along with the bracket defined above is a Leibniz algebra over A.*

Thus we have an infinitesimal deformation of the Leibniz algebra A which we denote by μ_L.

We have the following result.

Proposition 5.2. *Up to an isomorphism, this deformation does not depend on the choice of μ.*

Remark 5.3. In terms of a basis of H, here is an explicit description of the bracket giving μ_L. Suppose $\{h_1, h_2, \ldots, h_n\}$ is a basis of H and $\{t_1, t_2, \ldots, t_n\}$ is the dual basis. Under the identification, $A \otimes L = L \oplus Hom(H; L)$ an element $(l, \phi) \in L \oplus Hom(H; L)$ corresponds to $1 \otimes l + \sum_{i=1}^n t_i \otimes \phi(h_i)$. Let $\mu(h_i) = \mu_i \in CL^2(L; L)$. Then for $(l_1, \phi_1), (l_2, \phi_2)$ in $L \oplus Hom(H; L)$ their bracket $([l_1, l_2], \psi)$ corresponds to

$$1 \otimes [l_1, l_2] + \sum_{i=1}^{n} t_i \otimes (\mu_i(l_1, l_2) + [\phi_1(h_i), l_2] + [l_1, \phi_2(h_i)]).$$

The main property of μ_L is its co-universality in the class of infinitesimal deformations.

Theorem 5.4. *For any infinitesimal deformation l of the Leibniz algebra L with a finite dimensional base A, there exists a unique homomorphism*

$$\phi : \mathbb{K} \oplus HL^2(L; L)' \longrightarrow A$$

such that l is equivalent to the push out $\phi_* \mu_L$.

Sketch of the Proof. Let λ be an infinitesimal deformation of the Leibniz algebra L with a finite dimensional base A. Any $\xi \in \mathfrak{M}' = Hom_{\mathbb{K}}(\mathfrak{M}, \mathbb{K})$ can be regarded as an element in $A' = Hom_{\mathbb{K}}(A, \mathbb{K})$ with $\xi(1) = 0$. Then for any such ξ we have a cochain $\alpha_{\lambda,\xi} \in Hom(L^{\otimes 2}; L) = CL^2(L; L)$, defined by

$$\alpha_{\lambda,\xi}(l_1, l_2) = \xi \otimes id([1 \otimes l_1, 1 \otimes l_2]_\lambda) \in \mathbb{K} \otimes L = L.$$

Then we have

Lemma 5.5. *The cochain $\alpha_{\lambda,\xi} \in C^2(L; L)$ is a cocycle.*

Suppose $\{m_1, \ldots, m_k\}$ is a basis of $\mathfrak{M}$ and $\{\xi_1, \ldots, \xi_k\}$ be the corresponding dual basis of $\mathfrak{M}'$. For $\xi_i \in \mathfrak{M}'$ let $a_{\lambda,\xi_i} \in HL^2(L; L)$ be the cohomology class of the cocycle α_{λ,ξ_i}. The correspondence $\xi_i \longmapsto \alpha_{\lambda,\xi_i}$ and $\xi_i \longmapsto a_{\lambda,\xi_i}$ for $i = 1, \ldots, k$ will define homomorphisms $\alpha_\lambda : \mathfrak{M}' \longrightarrow C^2(L; L)$ with $\delta \circ \alpha_\lambda = 0$ and $a_\lambda : \mathfrak{M}' \longrightarrow HL^2(L; L)$. The main steps consist of proving the following two assertions.

(1) Two deformations λ_1 and λ_2 are equivalent if and only if $a_{\lambda_1} = a_{\lambda_2}$.
(2) For the map

$$\phi = id \oplus a_{\lambda'} : \mathbb{K} \oplus HL^2(L; L) \longrightarrow \mathbb{K} \oplus \mathfrak{M} = A$$

we have $\alpha_{\phi_* \mu_L} = \mu \circ a_\lambda$.

It is not possible to construct a formal deformation of a Leibniz algebra having a similar universal property. But it becomes possible by an appropriate modification of this property.

Definition 5.6. *A formal deformation η of a Leibniz algebra L with base B is called versal, if*

(i) *for any formal deformation λ of L with any (local) base A there exists a homomorphism $f : B \to A$ such that the deformation λ is equivalent to $f_* \eta$;*
(ii) *in the notations of (i), if A satisfies the condition $\mathfrak{M}^2 = 0$, then f is unique.*

Our goal is to construct a versal deformation of a given Leibniz algebra.

6. Obstruction to extending deformation

We need to use some results from Harrison cohomology of a commutative, associative algebra A over $\mathbb{K}$, with identity. We begin with the definition of Harrison cohomology of A with coefficients in a A-module M, denoted by $H^*_{Harr}(A; M)$.

The Harrison complex of A is induced from the Hochschild complex of A and is defined as follows.

Let S_n be the symmetric group of n symbols. Recall that a permutation $\sigma \in S_{p+q}$ is called a (p,q)-shuffle, if $\sigma(1) < \sigma(2) < \cdots < \sigma(p)$, and, $\sigma(p+1) < \sigma(p+2) < \cdots < \sigma(p+q)$. We denote the set of all (p,q)-shuffles in S_{p+q} by $Sh(p,q)$.

Consider $C_n(A) = A \otimes A^{\otimes n}$ where A operates on the first factor. If we denote the element $(a_0, a_1, \ldots, a_n) \in C_n(A)$ by $a_0(a_1, \ldots, a_n)$ and write $1(a_1, \ldots, a_n)$ simply by $(a_1, \ldots, a_n)$, then the boundary

$$\delta : C_n(A) \longrightarrow C_{n-1}(A)$$

is the A-linear map such that

$$\begin{aligned}\delta(a_1, a_2, \ldots, a_n) &= a_1(a_2, \ldots, a_n) + \sum_{i=1}^{n-1}(-1)^i(a_1, \ldots, a_i a_{i+1}, \ldots, a_n)\\ &\quad + (-1)^n a_n(a_1, \ldots, a_{n-1}). \end{aligned} \tag{4}$$

For $a_1, a_2, \ldots, a_n \in A$ and $0 < p < n$, set

$$s_p(a_1, a_2, \ldots, a_n) = \sum_{\sigma \in Sh(p,n-p)} sgn(\sigma)(a_{\sigma(1)}, a_{\sigma(2)}, \ldots, a_{\sigma(n)}) \in C_n(A),$$

where $Sh(p,q)$ is the set of all (p,q)-shuffles as defined above. Let $Sh_n(A)$ be the A-submodule of $C_n(A)$ generated by the chains $s_p(a_1, a_2, \ldots, a_n)$ for $a_1, a_2, \ldots, a_n \in A$ and $0 < p < n$.

It follows from Proposition 2.2 in [2] that $\delta(Sh_n(A)) \subset Sh_{n-1}(A)$. Thus we get a chain complex $Ch_*(A) = \{Ch_n(A), \delta\}$ where $Ch_n(A) = C_n(A)/Sh_n(A)$. This is known as the Harrison chain complex for the algebra A with trivial coefficients.

Let M be a left A-module. Since A is commutative, M can be considered as an A-bimodule. Consider the cochain complex $Ch^*(A; M)$ obtained by applying the functor $Hom(-; M)$ to the complex $Ch_*(A)$:

$$\begin{aligned}\cdots &\xrightarrow{\delta^{n-2}} Hom(Ch_{n-1}(A); M) \xrightarrow{\delta^{n-1}} Hom(Ch_n(A); M)\\ &\xrightarrow{\delta^n} Hom(Ch_{n+1}(A); M) \cdots\end{aligned}$$

Definition 6.1. *For an A-module M, the nth Harrison cohomology of A with coefficients in M is defined by*

$$H^n_{Harr}(A; M) = H^n(Ch^*(A; M)).$$

In our discussion we mainly require first and second Harrison cohomologies. In low dimension the Harrison cochain complex is given by

$$0 \longrightarrow Hom(A; M) \xrightarrow{\delta^1} Hom(S^2A; M) \xrightarrow{\delta^2} Hom(A^{\otimes 3}; M), \text{ where}$$

$$\delta^1\psi(a,b) = a\psi(b) - \psi(ab) + b\psi(a) \text{ for } \psi \in Hom(A; M)$$

$$\delta^2\phi(a,b,c) = a\phi(b,c) - \phi(ab,c) + \phi(a,bc) - c\phi(a,b)$$

$$\text{for } \phi \in Hom(S^2A; M) \text{ and } a,b,c \in A. \tag{5}$$

Remark 6.2. Note that $Ch_2(A)$ is by definition S^2A, the symmetric product module of A.

Definition 6.3. *An extension B of an algebra A by an A-module M is a $\mathbb{K}$-algebra B together with an exact sequence of $\mathbb{K}$-vector spaces*

$$0 \longrightarrow M \xrightarrow{i} B \xrightarrow{p} A \longrightarrow 0,$$

where p is a $\mathbb{K}$-algebra homomorphism, and the B-module structure on $i(M)$ is given by the A-module structure of M by $i(m) \cdot b = i(m \cdot p(b))$.

We have the following relationship between the Harrison cohomology and the extensions.

Proposition 6.4. *Elements of $H^2_{Harr}(A; M)$ correspond bijectively to isomorphism classes of extensions*

$$0 \longrightarrow M \longrightarrow B \longrightarrow A \longrightarrow 0$$

of the algebra A by M.

Let λ be a deformation of the Leibniz algebra L with a finite dimensional local base A. Consider $f \in H^2_{Harr}(A, \mathbb{K})$.

Suppose

$$0 \longrightarrow \mathbb{K} \xrightarrow{i} B \xrightarrow{p} A \longrightarrow 0$$

is a representative of the class of 1-dimensional extensions of A corresponding to the cohomology class f, where $\mathbb{K}$ is considered as an A-module via the augmentation $\varepsilon : A \longrightarrow \mathbb{K}$. Let $\imath = (i \otimes id) : L \cong \mathbb{K} \otimes L \longrightarrow B \otimes L,\ \ \pi = (p \otimes id) : B \otimes L \longrightarrow A \otimes L$ and $\eta = (\hat{\varepsilon} \otimes id) : B \otimes L \longrightarrow \mathbb{K} \otimes L \cong L$, where $\hat{\varepsilon} = \varepsilon \circ p$ is the augmentation for B. Fix a section $q : A \longrightarrow B$ of p in the above extension, then

$$b \longmapsto (p(b), i^{-1}(b - q \circ p(b)))$$

is a $\mathbb{K}$-module isomorphism $B \longrightarrow A \oplus \mathbb{K}$. Let us denote by $(a,k)_q \in B$ the inverse of $(a,k) \in A \oplus \mathbb{K}$ under the above isomorphism. The algebra structure of B determined by f is given by

$$(a_1, k_1)_q \circ (a_2, k_2)_q = (a_1a_2, a_1k_2 + a_2k_1 + f(a_1, a_2))_q$$

Now the Leibniz algebra structure λ on $A \otimes L$ can be lifted to a B-bilinear operation

$$\{\ ,\ \} : (B \otimes L) \times (B \otimes L) \longrightarrow B \otimes L$$

which satisfy the following

$$\text{(i)} \quad \pi\{l_1, l_2\} = [\pi(l_1), \pi(l_2)]_\lambda; \quad \forall\ l_1, l_2 \in B \otimes L$$
$$\text{(ii)} \quad \{\imath(l), l_1\} = \imath[l, \eta(l_1)];\ \forall l \in L \text{ and } l_1 \in B \otimes L.$$

In fact, suppose dim $(A) = r + 1$ and $\{m_1, \ldots, m_r\}$ be a basis of the maximal ideal $\mathfrak{M}_A$ of A. Then we can take $\{n_1, \ldots, n_{r+1}\}$ as a basis of the maximal ideal $\mathfrak{M}_B = p^{-1}(\mathfrak{M}_A)$ of B, where $n_j = (m_j, 0)_q$ for $j = 1, \ldots, r$ and $n_{r+1} = (0, 1)_q$. Let $\{m_1{}', \ldots, m_r{}'\}$ be the dual basis of $\mathfrak{M}_A{}'$. Then we have 2-cochains $\psi_i = \alpha_{\lambda, m_i'} \in CL^2(L; L)$ (cf. sketch of the proof of Theorem 5.4) for $i = 1, \ldots, r$ and $[\ ,\ ]_\lambda$ can be written as

$$[1 \otimes l_1, 1 \otimes l_2]_\lambda = 1 \otimes [l_1, l_2] + \sum_{i=1}^{r} m_i \otimes \psi_i(l_1, l_2); \quad \forall l_1, l_2 \in L.$$

Let $\psi \in CL^2(L; L) = Hom(L^{\otimes 2}; L)$ be an arbitrary element. Now if we define

$$\{\ ,\ \} : (B \otimes L)^2 \longrightarrow B \otimes L$$

by

$$\{b_1 \otimes l_1, b_2 \otimes l_2\} = b_1 b_2 \otimes [l_1, l_2] + \sum_{j=1}^{r} b_1 b_2 n_j \otimes \psi_j(l_1, l_2) + b_1 b_2 n_{r+1} \psi(l_1, l_2),$$

then it can be checked that $\{\ ,\ \}$ is a B-bilinear operation on $B \otimes L$ satisfying the conditions (i) and (ii) as mentioned above. For $x_1,\ x_2,\ x_3 \in B \otimes L$ define

$$\phi : (B \otimes L)^3 \longrightarrow B \otimes L$$

by

$$\phi(x_1, x_2, x_3) = \{x_1, \{x_2, x_3\}\} - \{\{x_1, x_2\}, x_3\} + \{\{x_1, x_3\}, x_2\}.$$

Observe that $\{\ ,\ \}$ satisfy Leibniz relation iff $\phi = 0$. Moreover, the map ϕ is multilinear and takes values in Ker π and $\phi(x_1, x_2, x_3) = 0$ if any one of $x_1,\ x_2$ or x_3 is from Ker η. This allows us to define a 3-cochain

$$\bar{\phi} : L^{\otimes 3} \cong (B \otimes L/\text{Ker } \eta)^{\otimes 3} \longrightarrow \text{Ker } \pi \cong L.$$

It turns out that $\bar{\phi}$ is a cocycle and the corresponding cohomology class $[\bar{\phi}]$ does not depend on the initial choice of the 2-cochain ψ, but only on the cohomology class $f \in H^2_{Harr}(A, \mathbb{K})$. This defines a linear map

$$\theta_\lambda : H^2_{Harr}(A; \mathbb{K}) \longrightarrow HL^3(L; L), \quad \theta_\lambda(f) = [\bar{\phi}].$$

We have the following criterion for extension.

Proposition 6.5. *The deformation λ with base A can be extended to a deformation of L with base B if and only if $\theta_\lambda(f) = 0$.*

Remark 6.6. The above results may be generalized from the case of extension $0 \to \mathbb{K} \to B \to A \to 0$ to a more general case of extension $0 \to M \xrightarrow{i} B \xrightarrow{p} A \to 0$, where M is a finite dimensional A-module satisfying the condition $\mathfrak{M}M = 0$. In this case, the obstruction would lie in $HL^3(L; M \otimes L) = M \otimes HL^3(L; L)$.

7. Construction of a versal deformation

We shall give a construction of a versal deformation starting with the infinitesimal universal deformation μ_L and using induction. We shall use a specific extension of a finite dimensional commutative local algebra A as described below. This is the extension,

$$0 \longrightarrow H^2_{Harr}(A; \mathbb{K})' \longrightarrow C \longrightarrow A \longrightarrow 0,$$

where the operation of A on $H^2_{Harr}(A; \mathbb{K})'$ is induced by the augmentation of A. The above extension of A is determined by the cocycle

$$f_A : S^2 A \longrightarrow H^2_{Harr}(A; \mathbb{K})'$$

which is defined as the dual of a homomorphism,

$$\mu : H^2_{Harr}(A; \mathbb{K}) \longrightarrow Ch^2(A; \mathbb{K}) = (S^2 A)'$$

where μ takes each cohomology class into a cocycle representing it (cf. Remark 6.2). This extension does not depend, up to isomorphism, on the choice of μ.

Now consider a Leibniz algebra L with dim $HL^2(L; L) < \infty$. $C_0 = \mathbb{K}$, $C_1 = \mathbb{K} \oplus HL^2(L; L)'$. Let

$$0 \longrightarrow HL^2(L; L)' \xrightarrow{i} C_1 \xrightarrow{p} C_0 \longrightarrow 0$$

be the canonical splitting extension, where the multiplication in C_1 is defined as

$$(k_1, h_1) \circ (k_2, h_2) = (k_1 \cdot k_2, k_1 \cdot h_2 + k_2 \cdot h_1);$$

for all $(k_1, h_1), (k_2, h_2) \in C_1$. Let η_1 denote the deformation of L with base C_1 which is the infinitesimal deformation μ_L as constructed in Section 5.

Remark 7.1. Algebra structure on C_1 depends only on the $\mathbb{K}$-vector space structure of $H^2(L; L)'$.

Suppose for some $k \geq 1$ we have constructed a finite dimensional commutative local algebra C_k and deformation η_k of L with base C_k. Let

$$\mu_k : H^2_{Harr}(C_k; \mathbb{K}) \longrightarrow (S^2 C_k)'$$

be the map sending a cohomology class to a representative of cocycle. Let

$$f_k : (S^2 C_k) \longrightarrow H^2_{Harr}(C_k; \mathbb{K})'$$

be the dual of μ. Then as discussed above, we get an extension of C_k

$$0 \to H^2_{Harr}(C_k; \mathbb{K})' \overset{\bar{i}_{k+1}}{\to} \bar{C}_{k+1} \overset{\bar{p}_{k+1}}{\to} C_k \to 0.$$

Now the Leibniz algebra structure $[\ ,\ ]_{\eta_k}$ on $C_k \otimes L$ can be lifted into a $\bar{C}_{k+1}$- bilinear map

$$\{\ ,\ \} : (\bar{C}_{k+1})^{\otimes 2} \longrightarrow (\bar{C}_{k+1})$$

which has the obstruction $\theta(f_k) \in H^2_{Harr}(C_k; \mathbb{K})' \otimes HL^3(L; L)$ by Remark 6.6. Now using the isomorphism

$$H^2_{Harr}(C_k; \mathbb{K})' \otimes HL^3(L; L) \cong \mathrm{Hom}(H^2_{Harr}(C_k; \mathbb{K}), HL^3(L; L))$$

$\theta(f_k)$ can be viewed as a linear map,

$$\omega_k : H^2_{Harr}(C_k; \mathbb{K}) \longrightarrow HL^3(L; L).$$

The dual map is given by

$$\omega_k{}' : HL^3(L; L)' \longrightarrow H^2_{Harr}(C_k; \mathbb{K})'.$$

Now set $C_{k+1} := \bar{C}_{k+1}/(\bar{i}_{k+1} \circ \omega_k'(HL^3(L; L)'))$. Then the above extension induces an extension

$$0 \to \text{Cokernel } \omega_k' \to C_{k+1} \to C_k \to 0,$$

and hence an extension

$$0 \to (\text{Ker } \omega_k)' \overset{i_{k+1}}{\to} C_{k+1} \overset{p'_{k+1}}{\to} C_k \to 0.$$

Notice that all the algebras C_k are local. This is because for any extension

$$0 \to M \overset{i}{\to} B \overset{p}{\to} A \to 0$$

of an algebra A, A is local with maximal ideal $\mathfrak{M}$ implies B is local with maximal ideal $p^{-1}(\mathfrak{M})$. Moreover, as C_k is finite dimensional, the cohomology group $H^2_{Harr}(C_k; \mathbb{K})$ is also finite dimensional and hence C_{k+1} is also finite dimensional. Now if we try to extend the deformation η_k of L from C_k to C_{k+1} then as discussed above the obstruction would be a homomorphism Ker $\omega_k \to HL^3(L; L)$. It turns out that this homomorphism is precisely the restriction of ω_k. Hence the obstruction is 0. This gives us the following result.

Proposition 7.2. *The deformation η_k admits an extension to a deformation η_{k+1} with base C_{k+1}, and this extension is unique up to an isomorphism of an extension,*

$$0 \longrightarrow (ker\ \omega_k)' \overset{i_{k+1}}{\longrightarrow} C_{k+1} \overset{p_{k+1}}{\longrightarrow} C_k \longrightarrow 0$$

By induction we get a sequence of finite dimensional local algebras C_k and deformations η_k of the Leibniz algebras L with base C_k,

$$\mathbb{K} \overset{p'_1}{\longleftarrow} C_1 \overset{p'_2}{\longleftarrow} C_2 \overset{p'_3}{\longleftarrow} \ldots\ldots \overset{p'_k}{\longleftarrow} C_k \overset{p'_{k+1}}{\longleftarrow} C_{k+1} \ldots,$$

such that

$$p'_{k+1*}\eta_{k+1} = \eta_k.$$

Thus by passing to the limit we obtain a formal deformation η of L with base $C = \varprojlim C_k$. The main result here is:

Theorem 7.3. *If* dim $(HL^2(L; L)) < \infty$, *then the formal deformation* η *constructed above is a versal deformation of* L.

See [4] for details.

Remark 7.4. The theory described above can be used to compute the versal deformation of specific examples. For detailed computations in the case of the Example 3.6, see [4], [20]. Since any Lie algebra is also a Leibniz algebra, it is natural to ask if one gets back only the Lie deformations of a Lie algebra when one deforms the Lie algebra as a Leibniz algebra. This question has been investigated in [4] for 3-dimensional complex Heisenberg group. It turns out that in this case, not only one recovers the known Lie deformations, but some new deformations which are Leibniz algebras also come up. There are many related interesting questions which need to be answered. For instance, we do not know if a complex semisimple Lie algebra is rigid as a Leibniz algebra.

References

[1] S. Albeverio, B. A. Omirov and I. S. Rakhimov, Varieties of Nilpotent complex Leibniz algebras of dimension less than five, *Communications in Algebra*, **33** (2005) 1575–1585.

[2] M. Barr, Harrison homology, Hochschild homology and triples, *J. Algebra*, **8** (1968) 314–323.

[3] A. Fialowski and Dmitry Fuchs, Construction of Miniversal deformations of Lie algebras, *Journal of Func. Anal.*, **161** (1967) 76–110.

[4] A. Fialowski, A. Mandal and G. Mukherjee, Versal Deformations of Leibniz algebras, *Journal of K-theory*, **3(2)**, (2009) 327–358 (arXiv:math.QA/0702476, 2008).

[5] A. Fialowski and M. Penkava, Versal Deformations of Three-dimensional Lie Algebras as L_∞ algebras, *Comm. Cont. Math.*, **7** (2005) 145–165.

[6] T. F. Fox, An introduction to algebraic deformation theory, *J. Pure Appl. Alg.*, **84** (1993) 17–41.

[7] M. Gerstenhaber, The cohomology structure of associative rings, *Ann. of Math.*, **78** (1963) 267–288.

[8] M. Gerstenhaber, On the deformations of rings and algebras, *Ann. of Math.*, **79** (1964) 59–103.

[9] M. Gerstenhaber, On the deformations of rings and algebras, *Ann. of Math.*, **84** (1966) 1–19.

[10] M. Gerstenhaber, On the deformations of rings and algebras, *Ann. of Math.*, **88** (1968) 1–34.

[11] M. Gerstenhaber, On the deformations of rings and algebras, *Ann. of Math.*, **99** (1974) 257–276.

[12] M. Gerstenhaber and S.D. Schack, Bialgebra Cohomology, Deformations and Quantum Groups, *Proc. Nat. Acad. Sci. USA*, **87** (1990) 478–481.

[13] K. Kodaira, L. Nirenberg and D. C. Spencer, On the Existence of Deformations of Complex Analytic Structures, *Ann. of Math.*, **68** (1958) 450–459.

[14] K. Kodaira and D. C. Spencer, A Theorem of Completeness for Complex Analytic Fibre Space, *Acta Math.*, **100** (1958) 281–294.

[15] K. Kodaira and D. C. Spencer, On Deformations of Complex Analytic Structures I-II, *Ann. of Math.*, **67** (1958) 328–466.

[16] J. L. Loday, Une version non-commutative des algebres de Lie: Les algebres de Leibniz, *Enseign. Math.*, **39** (1993) no. 3–4, 269–293.

[17] J. L. Loday, Overview on Leibniz algebras, dialgebras and their homology, *Fields Institute Communications*, **17** (1997) 91–102.

[18] J. L. Loday, Dialgebras and related operads, *Lecture Notes in Math.*, **1763** (2001) 7–66.

[19] A. Mandal, Deformation of Leibniz algebra morphisms, *Homology, Homotopy and Applications*, **9(1)** (2007) 439–450.

[20] A. Mandal, An example of constructing versal deformations for Leibniz algebras, *Comm. Algebra*, **36(6)** (2008) 2233–2253.

[21] Anita Majumdar and Goutam Mukherjee, Deformation theory of dialgebras, *K-Theory*, **27** (2002) 33–60.

[22] A. Nijenhuis and R. Richardson, Cohomology and deformation of graded Lie algebras, *Bull. Amer. Math. Soc.*, **72** (1966) 1–29.

[23] A. Nijenhuis and R. Richardson, Commutative algebra cohomology and deformation of Lie and associative algebras, *J. Algebras*, **9** (1968) 42–105.

[24] A. Nijenhuis and R. Richardson, Deformation of Lie algebra structures, *J. Math. Mech.*, **17** (1967) 89–105.

[25] W. S. Piper, Algebraic Deformation Theory, *J. Diff. Geo.*, **1** (1967) 133–168.

[26] D. C. Spencer, Deformations of Structures of Manifolds defined by Transitive, Continuous, Pseudographs I-II, *Ann. of Math.*, **76** (1962) 306–455.

[27] D. C. Spencer, Deformations of Structures of Manifolds defined by Transitive, Continuous, Pseudographs III, *Ann. of Math.*, **81** (1965) 389–450.

[28] M. Schlessinger, Functors of Artin rings, *Trans. Amer. Math. Soc.*, **130** (1968) 208–222.

[29] Donald Yau, Deformation theory of modules, *Comm. in Algebra*, **33** (2005) 2351–2359.

[30] Donald Yau, Deformation of dual Leibniz algebra morphisms, *Comm. Algebra*, **35(4)** (2007) 1369–1378.

[31] Donald Yau, Deformation theory of dialgebra morphisms, *Algebra Colloq.*, **15(2)** (2008) 279–292.

Perspectives in Geometry and Topology No. 11, 2010, pp. 67–71.

Bruhat-Tits Theory

M. S. Raghunathan

School of Mathematics, Tata Institute of Fundamental Research, Mumbai 400 005, India
e-mail: msr@math.tifr.res.in

Abstract. This is a brief account of Bruhat-Tits theory of semisimple p-adic groups. The basic ideas and concepts are described in a way somewhat different from that in the basic papers of Bruhat and Tits.

Keywords. p-adic groups, building.

1.1. This is an expository article on Bruhat-Tits theory for algebraic groups over a local field of characteristic 0. The theory was developed by Bruhat and Tits in their two papers [2] and [3].

Recall that a local field of characteristic 0 is a finite extension of the completion $\mathbb{Q}_p$ of $\mathbb{Q}$ with respect to the p-adic valuation, p a prime. In the sequel we fix such a field and denote it by k. Let $G \subset GL(n,k)$ (for some integer $n > 0$) be a (linear) algebraic group (over k) i.e. G is a subgroup of $GL(n,k)$ which is also the set of zeros of a finite number of functions on $GL(n,k)$ which are of the form $P_\alpha(X_{ij}, (\det X_{ij})^{-1})$ where $X_{ij} : GL(n,k) \to k,\ 1 \leq i,\ j \leq n$, is the function that assigns to each matrix X in $GL(n,k)$ its $(ij)^{-th}$ entry and $P_\alpha, \alpha \in S$, are polynomials in (n^2+1) variables with coefficients in k. In the sequel, if G is an algebraic group over k and k' is a finite extension of k, by the *group of k' points of G*, we mean the set (it is a group) of matrices in $GL(n,k')$ on which all the P_α vanish. The (algebraic) group G is *semisimple* iff it has no infinite abelian normal subgroup. In the sequel we fix a semisimple G. Bruhat-Tits theory among other things provides us with a locally compact contractible metric polyhedron on which the group G operates properly and this action enables one to prove many interesting properties of the group G. We give below a description of this metric space and briefly state some results (without proof) about them. Our formulations are not to be found in the available literature on the subject. The definitions below are equivalent to those of Bruhat-Tits but unlike those in the original work are made without the elaborate preliminaries needed in that fundamental paper. However to get the interesting properties from these definitions or even to make sure that the definitions are meaningful one seems to need the Bruhat-Tits machinery – we do not have any direct proofs.

1.2. Before we proceed further we make some further assumptions about G.

(i) We assume G to be "connected and simply connected". One interpretation of this is the following. Let $\bar{k}$ be an algebraic closure of k. Then k is isomorphic (as a field) to $\mathbb{C}$ the field of complex numbers. If we fix an isomorphism φ of $\bar{k}$ with

$\mathbb{C}$, then we may treat the polynomials $P_\alpha, \alpha \in S$ as polynomials over $\mathbb{C}$ and the set of zeros of the $P_\alpha(X_{ij}, \det(X_{ij})^{-1})$ in $GL(n, \mathbb{C})$ is a closed subgroup say $\bar{G}$ of $GL(n, \mathbb{C})$ which inherits a (Hausdorff) topology from that on $GL(n, \mathbb{C})$. Then G is connected and simply connected iff $\bar{G}$ as a topological space is (arcwise) connected and simply connected. A priori this seems to depend on the choice of φ but it is independent of this choice.

(ii) We assume that $\bar{G}(\subset GL(n, \mathbb{C}))$ has no proper infinite normal subgroups.

Once again this condition is independent of the choice of φ. We will say that G is absolutely almost simple if it has the above property. It turns out that if G is connected, simply connected and semisimple, then G is a finite direct product $\prod_{i \in I} G_i$ where for each i in I, $G_i \subset G$ is a normal subgroup which is algebraic semisimple, absolutely almost simple and simply connected over a suitable finite extension k_i of k. We describe the Bruhat-Tits complex in the case of an absolutely almost simple group; in the general case the group being a product of absolutely almost simple groups, the Bruhat-Tits space is simply the product of the spaces corresponding to the different absolutely almost simple factors.

1.3. The field k has a locally compact topology which in turn enables one to give a natural (Hausdorff) topology on $GL(n, k)$ and hence on G making G into a locally compact group. If $\mathcal{O}$ is the ring of integers in k, $GL(n, \mathcal{O})$ is a (maximal) compact open subgroup of $GL(n, K)$. If $\mathbf{p}$ denotes the prime ideal in $\mathcal{O}$ and if we set $M = \{g \in GL(n, \mathcal{O}) | g \text{ is upper triangular unipotent mod } \mathbf{p})\}$, then it is easy to see that M is a pro-nilpotent pro-finite (open compact) subgroup of $GL(n, k)$. It is in fact a *maximal* pro-$\mathbf{p}$ subgroup of $GL(n, k)$. It is easy to see also that all maximal pro-$\mathbf{p}$ subgroups of $GL(n, k)$ are conjugate to M and that any compact subgroup of $GL(n, k)$ is contained in a conjugate of $GL(n, \mathcal{O})$ and hence $GL(n, \mathcal{O})$ is a maximal compact (open) subgroup of $GL(n, k)$.

Definition 1.4. *An Iwahori subgroup of G is the normalizer of a maximal pro-$\mathbf{p}$ subgroup of G.*

1.5. This definition make sense because of the following facts (which require some work to prove) which are analogues of the Sylow theorem for finite groups.

(i) Every pro-$\mathbf{p}$-subgroup of G is contained in a maximal pro-$\mathbf{p}$-subgroup of G.
(ii) Any two maximal pro-p-subgroups of G are conjugate by an element of G.

Thus the Iwahori subgroup is *unique up to conjugacy.*

1.6. A *split torus* in G is an algebraic subgroup T of G which is isomorphic to a finite direct product of copies of k^*(and can be conjugated into diagonal matrices in $GL(n, k)$ by an element of $GL(n, k)$). Evidently any split torus in G is contained in a maximal one. A theorem of Borel-Tits assures us that all maximal k-split tori G are conjugates of one another by elements of G. Thus the dimensions of a maximal k-split tori (as Lie groups over k) depends only on G and is called the k-rank of G. We fix a maximal split torus T in G once and for all. Now it is not difficult to see that any split torus in G has a *unique* maximal compact sub group; this holds in particular for T. We denote by U the maximal compact subgroup in T. Now T is isomorphic to a product of copies of k^* and each factor has a unique maximal compact subgroup $\mathcal{O}^*$-the units in the ring $\mathcal{O}$. We denote $\mathcal{O}_1^*$ the subgroup $\{x \in \mathcal{O}^* | x \equiv 1 \bmod \mathbf{p}\}$; then $\mathcal{O}_1^*$ is a maximal prop-p-group in k^*. Taking the product of the $\mathcal{O}_1^*$ corresponding to the various factors of T, we obtain a maximal pro-p-subgroup U_1 of T (contained in U). Now let I be an Iwahori subgroup containing U_1. With these definitions we have the following.

Theorem 1.7. *Let N denote the normalizer of T in G. Then N normalizes U as well as U_1 and N/U has a system of involutive generators S such that (G, I, N, S) is a Tits system.*

That (G, I, N, S) is a Tits system means the following: I and N together generate G, $N \cap I = U$ is normal in N and $(N/N \cap I, S)$ is a Coxeter system; for $s \in S$ and $w \in W(= N/N \cap I)$, one has $s.\, IwI = IwI \cup I\, sw\, I$, where we have let w stand for any lift of w to N. (cf. [1].)

1.8. We are now in a position to describe the Bruhat-Tits complex, but first a definition. A subgroup $P \subset G$ is a *parahoric* subgroup if P contains an Iwahori subgroup. The Bruhat-Tits complex is the simplicial complex obtained as follows: its vertices are proper maximal parahoric subgroups of G. A set $(P_0, P_1, \ldots, P_r)$ of vertices is a $(r-)$ simplex if $P_0 \cap P_1 \cap \cdots \cap P_r$ is a parahoric subgroup. We denote the geometric realization of this simplicial complex by $\mathbf{B}$. Evidently G acts on $\mathbf{B}$ by inner conjugation.

1.9. From standard results about Tits-systems one has the following facts:

- If $S' \subset S$ is a subset, $W(S')$ the subgroup generated by S' in N/U and $\widehat{W}(S')$ the inverse image of $W(S')$ in N, then $I\widehat{W}(S')I$ is a parahoric subgroup $P(S')$ of G.
- Every parahoric subgroup P is conjugate to a $P(S')$ for a suitable subset S' of S; moreover S' is uniquely determined by P (in particular $P(S')$ is not conjugate to $P(S'')$ if $S' \neq S''$, $S', S'' \subset S$).

1.10. $\mathbf{B}$ is a locally compact space and the action of G on $\mathbf{B}$ is proper. One introduces a metric on $\mathbf{B}$. For this we need to collect together a number of facts first. The group G acts on the collection of all parahoric subgroups (by inner conjugation). The sub collection of all Iwahori subgroups is stable under this action; further G acts transitively on this set. One sees immediately that this action yields a *simplicial* action of G on $\mathbf{B}$. The isotropy at a parahoric subgroup for the action on the set of parahoric groups is the parahoric subgroup itself (and a parahoric group is its own normalizer). The parahoric subgroups correspond to simplices of $\mathbf{B}$; this correspondence is the one that assigns to each simplex of $\mathbf{B}$, the stabilizer of the simplex for the natural action on $\mathbf{B}$. The dimension of $\mathbf{B}$ is the k-rank of G and the maximal dimensional simplices correspond bijectively to Iwahori subgroups. The group G acts transitively on the set of maximal dimensional simplices. Each simplex is contained in a maximal dimensional simplex. If we make the standard identification of a maximal dimensional simplex with $\{\underline{x} \in \mathbb{R}^{r+1} | x_i \geq 0; \Sigma x_i = 1\}$, we have on each maximal dimensional simplex (and hence on any simplex) a metric: the one deduced from the metric on $\mathbb{R}^{r+1}$. (Here r= rank G.) With this background information the metric d on **B** is defined as follows: A piecewise linear path (pl-path) in **B** is a path $\gamma : [0, 1] \to \mathbf{B}$ such that there are real numbers $t_0, t_1, \ldots, t_q$ (for some integer $q \geq 0$) such that $0 = t_0 < t_1 < t_2 < \cdots < t_q = 1$, $\gamma([t_i, t_{i+1}])$ is in a simplex and $\gamma|_{[t_i,t_{i+1}]}$ is affine linear. We denote (for γ as above) by $l(\gamma)$ the number $\Sigma_{0 \leq i < q} l_i$ where l_i is the distance between $\gamma(t_i)$ and $\gamma(t_{i+1})$ in the simplex in which the two points lie; then for $p, q \in \mathbf{B}$,

$$d(p, q) = inf\{l(\gamma)|\gamma \text{ is a pl-path in } \mathbf{B} \text{ with } \gamma(0) = p, \gamma(1) = q\}.$$

1.11. The space **B** is complete in the metric defined in 1.10. It has many of the properties that non-positively curved complete Riemannian manifolds have. We list some of them below.

(i) **B** is contractible.

(ii) Given $(p, q) \in \mathbf{B}$, there is up to piecewise linear reparameterization a unique pl-path $\gamma : [0, 1] \to \mathbf{B}$ such that $\gamma(0) = p, \gamma(1) = q$ and $d(p, q) = l(\gamma)$. Such γ are the "geodesics" in **B**.

(iii) Let T be a maximal k-split torus and U its maximal compact subgroup. Let $A(T)$ be the "full" subcomplex of **B** whose vertices are the maximal parahoric subgroups containing U. Then $A(T)$ with the metric induced from **B** is isometric to the Euclidean space $\mathbb{R}^r$. $A(T)$ is the "apartment" associated to T. Central to the theory is the following fixed point theorem.

Theorem 1.12. *Let H be a compact group acting on* **B** *as isometries. Then H has a fixed point in* **B**.

Corollary 1.13. *Any compact subgroup of G is contained in a maximal parahoric subgroup (and hence maximal compact subgroups of G exist and are the same as maximal parahoric subgroups).*

Corollary 1.14. *The maximal parahoric subgroups of G are the isotropy groups of vertices of* **B**. *If σ is a maximal $(r-)$ dimensional simplex in* **B**, *then the isotropy groups of distinct vertices of σ are not conjugate to each other. In particular there are exactly $(r + 1)$ distinct conjugacy classes of maximal compact subgroups in G.*

1.15. We have given above an account of the Bruhat-Tits theory which we hope gives a reasonably clear picture of what the basic geometric facts in the theory are. We have of course given no proofs. There are in fact no direct proofs in the literature of the results we have stated starting from our definitions. The theory as developed by Bruhat-Tits follows a different trajectory: what they do is first develop the entire theory for the special case of the so called quasi-split groups i.e. the groups for which the centralizer of a k-split torus is a maximal torus). The definition of the Iwahori subgroup is very different and is dependent on the concept of a "root-datum". For a general group G, they use the fact that G becomes quasi split over a maximal *unramified* extension $\hat{k}$ of k. Then one fixes a minimal cyclic extension k' of k such that k'-rank $G = \hat{k}$-rank G and study the action of Gal k'/k on the Bruhat-Tits building of G', the k'-points of G. The theory is already developed for G' and in particular one can apply the central fixed point theorem to the Galois action. They construct the Bruhat-Tits building for G as the fixed point set for this Galois action and develop the theory for G itself. That our definition of **B** coincides with theirs can be established using the theory so developed.

1.16. Among the other things they prove the following beautiful result.

Theorem 1.17. *Let P be a parahoric subgroup of G. Let $k[G]$ denote the coordinate ring of G and let $\mathcal{O}_P[G]$ be the sub-ring $\{f \in k[G] | f(P) \subset \mathcal{O}\}$. Then $\mathcal{O}_P[G]$ is a finitely generated $\mathcal{O}$-algebra and Spec $\mathcal{O}_P[G]$ is a smooth group scheme over $\mathcal{O}$ with $\mathcal{O}_P[G] \otimes_{\mathcal{O}} k \simeq k[G]$. Moreover, P is precisely the group of $\mathcal{O}$-points of the scheme Spec $\mathcal{O}_P[G]$.*

Bruhat-Tits also give a nice description of the (smooth) group scheme $\mathcal{O}_P[G] \otimes_{\mathcal{O}} F$ where F is the residue field of k.

1.18. We end this account with the description of the so called Iwasawa decomposition. Let $\mathcal{O}$ be as above. Let T be a maximal k-split torus. Then there is a maximal unipotent subgroup N of G normalized by T and such that the Lie subalgebra of Lie G (the Lie algebra of G) is spanned by eigen spaces of T (acting on Lie G by adjoint action). Then there is a maximal compact subgroup K of G such that $G = K.T.N$. This decomposition holds only for a maximal compact subgroup chosen in suitable conjugacy classes, not for all.

As has already been remarked this account attempts to outline some of the salient points in the Bruhat-Tits theory with minimal back-ground prerequisites; but this alternate description by no means leads to any simple proofs of the main results of the theory. It is hoped that it will be of some help in understanding the theory as developed by Bruhat-Tits in their great fundamental papers.

References

[1] N. Bourbaki, Elements of Mathematics, Lie Groups and Lie Algebras, Chapters 4–6, Springer-Veralg, Berlin-Hedelberg (2002) 15–26.

[2] F. Bruhat and J. Tits, Groupes reductifs sur un corps local I, Publications Mathematiques IHES 41 (1972) 5–251.

[3] F. Bruhat and J. Tits, Groupes reductifs sur un corps local II. Schemas en groupes. Existence d'une donneé radicielle valueé, Publications Mathematiques IHES 60 (1984) 197–376.

Perspectives in Geometry and Topology No. 11, 2010, pp. 73–91.

Lectures on the Wall's Finiteness Obstruction

K. Varadarajan

University of Calgary, Canada
e-mail: varadara@math.ucalgary.ca

Abstract. If X is dominated by a finite complex L, is X homotopically equivalent to a finite complex? This long standing question of J.H.C. Whitehead was answered in the negative in a seminal paper by C.T.C. Wall. In this survey article, we present an outline of Wall's result and some later related results on nilpotent spaces.

Keywords. CW-complexes, finite domination, nilpotent spaces, ideal class group.

1. Weak homotopy equivalences

For any integer $k \geq 1$, $\mathbb{R}^k$ denotes the real Euclidean space of dimension k with the usual metric topology. The unit disk $\{(x_1, \ldots, x_k) \in \mathbb{R}^k | \sum_{i=1}^{k} x_i^2 \leq 1\}$ in $\mathbb{R}^k$ is denoted by E^k and the unit sphere $\{(x_1, \ldots, x_k) \in \mathbb{R}^k | \sum_{i=1}^{k} x_i^2 = 1\}$ in $\mathbb{R}^k$ is denoted by S^{k-1}. The space consisting of one element $\{0\}$ is denoted by $\mathbb{R}^\circ$ and E° is the same as $\mathbb{R}^\circ$. By definition, the sphere S^{-1} of dimension-1 is the empty set $\emptyset$. By a topological pair (X, A) we mean a pair with X a topological space and A a subspace of X. Maps between topological spaces will be assumed to be continuous.

Definition 1.1. *Let (X, A) be a topological pair with $A \neq \emptyset$ and n an integer ≥ 0. We say that (X, A) is n-connected if for every $k \leq n$, every map $\alpha : (E^k, S^{k-1}) \to (X, A)$ is homotopic relative to S^{k-1} to a map $\beta : (E^k, S^{k-1}) \to (X, A)$ satisfying $\beta(E^k) \subseteq A$.*

Note that (X, A) is 0-connected iff every point in X can be joined by means of an arc in X to a point in A, equivalently each arcwise connected component of X has a non-empty intersection with A.

For $k \geq 1$, the point $(1, 0, 0, \ldots, 0) \in S^{k-1}$ will be denoted by p_0. Let $x_0 \in A$ and $\pi_k(X, A, x_0)$ the set of homotopy classes of maps $(E^k, S^{k-1}, p_0) \to (X, A, x_0)$. It is a group for $k \geq 2$ and a set with a distinguished 'zero' element when $k = 1$. The following is a well-known result in homotopy theory.

Proposition 1.2. *Let $k \geq 1$ and $\alpha : (E^k, S^{k-1}, p_0) \to (X, A, x_0)$ a given map. Then $[\alpha] = 0$ in $\pi_k(X, A, x_0)$ if and only if α is homotopic relative to S^{k-1} to a map $\beta : (E^k, S^{k-1}, p_0) \to (X, A, x_0)$ satisfying $\beta(E^k) \subseteq A$.*

For a proof see Theorem 1, Chapter 7, Section 2 of [7].

Corollary 1.3. *Let n be an integer ≥ 1. Then (X, A) is n-connected if and only if each arcwise connected component of X intersects A and $\pi_k(X, A, a) = 0$ for each $a \in A$ and $1 \leq k \leq n$.*

Remark 1.4. Suppose $A = \{x_0\}$. Then (X, A) is 0-connected if and only if X is arcwise connected. For $n \geq 1$, (X, A) is n-connected if and only if $\pi_k(X, x_0) = 0$ for $1 \leq k \leq n$. However, when X is arcwise connected $\pi_k(X, a) \approx \pi_k(X, b)$ for any two elements a, b of X. We can write the condition $\pi_k(X, x_0) = 0$ as $\pi_k(X) = 0$. This motivates the following.

Definition 1.5. *Let X be a non-empty topological space. We say that X is 0-connected if it is arcwise connected. For $n \geq 1$ we say that X is n-connected if X is 0-connected and $\pi_k(X) = 0$ for $1 \leq k \leq n$.*

Let X, Y be non-empty topological spaces and $f : X \to Y$ a map.

Definition 1.6. *The mapping cylinder of f, denoted by M_f is the quotient space obtained from the disjoint union $X \times I \dot{\cup} Y$ by identifying $(x, 1) \in X \times I$ with $f(x)$ for every $x \in X$.*

Writing $[x, t]$ for the element in M_f represented by (x, t) in $X \times I$, we can consider X as a subspace of M_f via $x \overset{j}{\mapsto} [x, 0]$. Writing $\bar{y}$, for the element of M_f represented by $y \in Y$, we can regard Y as a subspace of M_f via the map $\mu : Y \to M_f$ given by $\mu(y) = \bar{y}$. Actually Y is a strong deformation retract of M_f. The map $\nu : M_f \to Y$ given by $\nu([x, t]) = \overline{f(x)}$ and $\nu(\bar{y}) = y$ realizes Y as a strong deformation retract of M_f. Also $\nu \circ j(x) = f(x)$ for all $x \in X$.

Definition 1.7. *Let n be an integer ≥ 0. We say that a map $f : X \to Y$ is n-connected if (M_f, X) is n-connected.*

For any topological space X, let $\pi_0(X)$ denote the set of arc components of X. The map $f : X \to Y$ induces a function $f_* : \pi_0(X) \to \pi_0(Y)$ defined by $f_*(C) =$ the path component of Y containing $f(C)$, for any path component C of X. It is easily checked that $f : X \to Y$ is 0-connected if and only if $f_* : \pi_0(X) \to \pi_0(Y)$ is onto.

Proposition 1.8. *Let n be an integer ≥ 1. Then $f : X \to Y$ is n-connected if and only if $f_* : \pi_0(X) \to \pi_0(Y)$ is a bijection and $f_* : \pi_k(X, x) \to \pi_k(Y, f(x))$ is an isomorphism for $1 \leq k < n$ and onto for $k = n$, whatever be $x \in X$.*

Proof. Immediate consequence of the homotopy exact sequence of the pair (M_f, X).

Definition 1.9. *We refer to $f : X \to Y$ as a weak equivalence if it is n-connected for every integer $n \geq 1$.*

As an immediate consequence of Proposition 1.8 we obtain the following Corollary.

Corollary 1.10. *A map $f : X \to Y$ is a weak equivalence if and only if $f_* : \pi_0(X) \to \pi_0(Y)$ is a bijection and $f_* : \pi_k(X, x) \to \pi_k(Y, f(x))$ is an isomorphism for all integer $k \geq 1$ and all $x \in X$.*

A particular case of Corollary 1.10 is the following

Corollary 1.11. *Let $f : X \to Y$ be a map of 0-connected spaces, $x_0 \in X$ and $y_0 = f(x_0) \in Y$. Then f is a weak equivalence if and only if $f_* : \pi_k(X, x_0) \to \pi_k(Y, y_0)$ is an isomorphism for all $k \geq 1$.*

An important result due to J.H.C. Whitehead is the following:

Theorem 1.12 (J.H.C. Whitehead). *Let $f : X \to Y$ be a map of 0-connected spaces and n an integer ≥ 1.*

(i) *If f is n-connected, then $f_* : H_k(X) \to H_k(Y)$ is an isomorphism for $k < n$ and onto for $k = n$.*
(ii) *If X, Y are 1-connected and $f_* : H_k(X) \to H_k(Y)$ is an isomorphism for $k < n$ and onto for $k = n$ then f is n-connected.*
(iii) *If f is weak equivalence, $f_* : H_k(X) \to H_k(Y)$ is an isomorphism for all k.*
(iv) *If X, Y are 1-connected and $f_* : H_k(X) \to H_k(Y)$ is an isomorphism for all k then f is a weak equivalence.*

Theorem 1.12 follows from the relative Hurewicz Theorem. For more details please refer to §1, Chapter I of [10].

2. CW-complexes

Let Y be a topological space, E_λ a disk of dimension n_λ for λ in some indexing set J. Let $\dot{E}_\lambda = S_\lambda^{n_\lambda - 1}$ be the boundary sphere of the disk E_λ and $f_\lambda : S_\lambda^{n_\lambda - 1} \to Y$ be given maps. The quotient space obtained from the disjoint union $Y\dot{\cup}(\dot{\cup}_{\lambda\in J} E_\lambda^{n_\lambda})$ by identifying $a \in S_\lambda^{n_\lambda - 1}$ with $f_\lambda(a)$ in Y is denoted by $Y \cup_{f_\lambda} (\cup e_\lambda)$. Here $e_\lambda = \eta(E_\lambda)$ where $\eta : Y\dot{\cup}(\dot{\cup} E_\lambda^{n_\lambda}) \to Y\cup_{f_\lambda}(\cup e_\lambda)$ is the quotient map. Since $a \in S_\lambda^{n_\lambda - 1}$ is identified with $f_\lambda(a)$ we see that $\eta|E_\lambda^{n_\lambda} : E_\lambda^{n_\lambda - 1} \to Y \cup_{f_\lambda} (\cup e_\lambda)$ is an extension of $f_\lambda : S_\lambda^{n_\lambda - 1} \to Y$. We write $\dot{e}_\lambda$ for $f_\lambda(\dot{E}_\lambda) = f_\lambda(S_\lambda^{n_\lambda - 1})$. The e_λ are called the closed cells of $Y \cup_{f_\lambda} (\cup e_\lambda)$. If $g_\lambda = \eta|E_\lambda$ then $g_\lambda(E_\lambda) = e_\lambda$, $g_\lambda(\dot{E}_\lambda) = \dot{e}_\lambda$ and $g_\lambda : (E_\lambda, \dot{E}_\lambda) \to (e_\lambda, \dot{e}_\lambda)$ is a relative homeomorphism, namely, $g_\lambda : E_\lambda \backslash \dot{E}_\lambda \to e_\lambda \backslash \dot{e}_\lambda$ is a homeomorphism.

Definition 2.1. *The space $Y \cup_{f_\lambda} (\cup e_\lambda)$ is said to be obtained from Y by attaching the family of cells $(E_\lambda)_{\lambda\in J}$ by means of the maps $f_\lambda : S_\lambda^{n_\lambda - 1} \to Y$. The map g_λ is referred to as the characteristic map for the cell e_λ.*

If Y is Hausdorff, it can be shown that $Y\cup_{f_\lambda}(\cup e_\lambda)$ is also Hausdorff.

Definition 2.2. *A CW-complex is a topological space X together with a sequence $X_0 \subset X_1 \subset X_2 \subset \ldots$ of closed subspaces X_n for $n \geq 0$ satisfying the following conditions:*

(i) $X = \cup_{n\geq 0} X_n$.
(ii) *X_0 is discrete.*
(iii) *For each $n \geq 1$, X_n is obtained from X_{n-1} by attaching a family of n-dimensional cells.*
(iv) *The topology of X is the "coherent topology" determined by the X_n's. This means that a subset C of X is closed in X if and only if $C \cap X_n$ is closed in X_n for each $n \geq 0$.*

We state without proofs some properties of CW-complexes.

2.3. Let $(e^n_\alpha)_{\alpha \in J_n}$, $n \geq 0$, denote the closed cells of X. Then a set C is closed in X if and only if $C \cap e^n_\alpha$ is closed in e^n_α for every $\alpha \in J_n$ and $n \geq 0$. This property is usually stated as follows: The topology of X is the weak topology (or the coherent topology) determined by the closed cells of X.

2.4. X is Hausdorff, (in fact, paracompact.)

Let X be a Hausdorff space. Suppose there exist onto continuous maps $f^n_\alpha : E^n_\alpha \to e^n_\alpha$ with $e^n_\alpha \subset X, \alpha \in J_n$ and $n \geq 0$ where E^n_α is an n-disk satisfying the following conditions:

(1) Writing $\dot{e}^n_\alpha$ for $f_\alpha(S^{n-1}_\alpha)$ we require that $f_\alpha : (E^n_\alpha, S^{n-1}_\alpha) \to (e^n_\alpha, \dot{e}^n_\alpha)$ be a relative homeomorphism
(2) If $X_n = \bigcup_{0 \leq k \leq n} \bigcup_{\alpha \in J_k} e^k_\alpha$ then $\dot{e}^n_\alpha \subset X_{n-1}$ (where we set $X_{-1} = \phi$).
(3) Writing Int e^n_α for $e^n_\alpha \backslash \dot{e}^n_\alpha$ we require that $X = \dot{\bigcup}_{\alpha \in J_n, n \geq 0}$ Int e^n_α, a disjoint union.

In this case X is referred to as a cell complex.

The e^n_α's provide the cell structure of X and are called the closed cells of X.

A sub complex Y of X is a cell complex such that each (closed) cell of Y is a cell of X.

We again state without proofs the following properties of CW-complexes.

2.5. Any sub complex of a CW-complex X is a CW-complex.

2.6. Given any cell e of a CW-complex X, there exists a finite sub complex K of X with $e \subset K$.

A CW-complex X can alternatively be defined as a cell complex satisfying (2.3) and (2.6). Property (2.6) is referred to as closure finiteness.

Definition 2.7. *By a CW-pair (X, Y) we mean a pair where X is a CW-complex and Y a sub complex of X.*

Definition 2.8. *A CW-complex X is said to be of dimension $\leq n$ if $X = X_n$ (equivalently any cell of X is of dimension $\leq n$). A CW-pair (X, Y) is said to be of dimension $\leq n$ if any cell of X which is not a cell of Y is of dimension $\leq n$.*

We now state another important result due to J.H.C. Whitehead.

Theorem 2.9. *A map $f : X \to Y$ between CW-complexes is a homotopy equivalence if and only if it is a weak homotopy equivalence.*

A complete proof of this result is given in §2 Chapter I of [10].

3. Spaces dominated by CW-complexes

Definition 3.1. *A space X is said to be dominated by another space Y in homotopy if there exist maps $f : X \to Y$, $g : Y \to X$ satisfying $g \circ f \simeq Id_X$.*

The following result is due to J.H.C. Whitehead.

Theorem 3.2. *Let X be a space dominated by a CW-complex. Then X is of the homotopy type of a CW-complex.*

A complete proof of this result is given in §3, Chapter I of [10]. This proof uses Milnor's geometric realization of the singular semi-simplicial complex of X.

Definition 3.3. *A map $f : X \to Y$ of topological spaces will be called a homology equivalence if $f_* : H_k(X) \to H_k(Y)$ is an isomorphism for all $k \geq 0$.*

Theorem 2.9 and Theorem 3.2 yield the following.

Corollary 3.4. *Let X, Y be spaces dominated by CW-complexes. Then any weak homotopy equivalence $f : X \to Y$ is a homotopy equivalence.*

Also (iii), (iv) of Theorem 1.12 and Theorem 3.2 yield the following.

Corollary 3.5. *Let X, Y be 1-connected spaces dominated by CW-complexes. Then $f : X \to Y$ is a homotopy equivalence if and only if it is a homology equivalence.*

If X is dominated by a finite complex L, is X homotopically equivalent to a finite complex? J.H.C. Whitehead considered this to be a very difficult problem [12] and this problem remained unsolved for a long time till C.T.C. Wall gave a negative solution to this problem in his very original paper [11]. In this course of lectures we will give an account of this work of C.T.C. Wall. Later on in this course we will define nilpotent spaces. When X is simply connected and dominated by a finite complex, J.W. Milnor (unpublished) has shown that X is of the homotopy type of a finite complex. In [2] E. Dror obtained far reaching generalizations of J.R. Stallings's results on homotopy and central series of groups [8]. As a consequene of his results he proved the following result which extends the validity of J.H.C. Whitehead's result (Theorem 1.12 (iv) of the present notes) to the class of nilpotent spaces.

Theorem 3.6 (E. Dror). *Let $f : X \to Y$ be a map of 0-connected nilpotent spaces. If f is a homology isomorphism then it is a weak equivalence.*

This naturally suggests the question whether Milnor's result can be extended to the case of finitely dominated nilpotent spaces. G.Mislin answers this question in the negative, namely he shows that there exist finitely dominated nilpotent spaces which are not homotopically equivalent to finite complexes.

While we will give more details in our later talks on Wall's work we will give a brief description of the method used by him. With each 0-connected CW-complex X dominated by a finite complex, Wall associates an element $\tilde{\omega}(X) \in \tilde{K}_0(\mathbb{Z}\pi_1(X))$ which depends only on the homotopy type of X. Here $\tilde{K}_0(\mathbb{Z}\pi_1(X))$ is the reduced projective class group of the integral group ring $\mathbb{Z}\pi_1(X)$ of $\pi_1(X)$. He shows that X is of the homotopy type of a finite complex if and only if $\tilde{\omega}(X) = 0$. Given any finitely presented group π and any element $x \in \tilde{K}_0(\mathbb{Z}\pi)$, Wall shows that there exists a finitely dominated 0-connected CW-complex X with $\pi_1(X) \approx \pi$ and $\tilde{\omega}(X) = x$. The element $\tilde{\omega}(X)$ is called the (reduced) Wall obstruction of X. Mislin and to a lesser extent myself have carried out a thorough study of the Wall obstruction of finitely dominated nilpotent spaces. In these lectures I will present some of these results.

All the results stated by us in sections 1 and 2 and the results in section 3 until corollary 3.5 can be found in Whitehead's papers [12,13]. There is a good account of the results stated in sections 1 and 2 in Spanier's book [7]. In addition to the above sources, there is a good account in [10]. The results of E. Dror in [2] specialized to the case of a $K(\pi, 1)$ space will yield the results of J.R. Stallings [8] on homology and central series of groups.

4. CW-complexes of finite type and Milnor's construction

All the spaces considered in now onwards be 0-connected. The only exception will be when we consider the 0-skeleton of a CW-complex.

Let X be a space admitting a universal covering. We denote the universal covering of X by $\widetilde{X}$. However, there may be some occasions when for a subspace A of X we will write $\widetilde{A}$ for $p^{-1}(A)$ where $p : \widetilde{X} \to X$ is the covering projection. In such a situation, $\widetilde{A}$ in general will not be the universal covering of A. When we deal with such a situation we will explicitly mention it. For any CW-complex K and any integer $k \geq 0$, we write K^k for the k-skeleton of K. We will also write $K = K^k$ to mean that $\dim K \leq k$. (When $K = K^k$, it is not necessary that $\dim K$ is actually k. It can be $< k$). It will be assumed that all the spaces considered are of the homotopy type of a CW-complex.

A general construction originally due to John Milnor will be repeatedly used in our study of finiteness conditions for CW-complexes.

All the results presented in this section and the following one are due to C.T.C. Wall [11].

Definition 4.1. *A CW-complex K is said to be of finite type if it has finitely many cells in each dimension $n \geq 0$ or equivalently K^n is a finite complex for each integer $n \geq 0$.*

For any map $\phi : K \to X$ defines $\pi_n(\phi) := \pi_n(M_\phi, K)$, and $H_n(\phi) := H_n(M_\phi, K)$, for $n \geq 1$. Alternatively, following Eckmann and Hilton, $\pi_n(\phi)$ is the set of all homotopy classes of commutative diagrams

$$\begin{array}{ccc} S^{n-1} & \xrightarrow{f} & K \\ \downarrow & & \downarrow \phi \\ E^n & \xrightarrow{g} & X \end{array}$$

Thus ϕ is n-connected as defined in §1 if and only if $\pi_k(\phi) = 0$ for $1 \leq k \leq n$. Note that according to our convention K and X are 0-connected. However, S^0 is not 0-connected.

Let $K \xrightarrow{\phi} X$ be $(n-1)$ connected with $n \geq 2$. If $n \geq 3$ this implies that $\phi_* : \pi_1(K) \to \pi_1(X)$ is an isomorphism, and there exists a lift $\tilde{\phi} : \tilde{K} \to \tilde{X}$. In this case $\pi_n(\tilde{\phi}) \approx \pi_n(\phi)$ is a module over $\mathbb{Z}\pi$ where $\pi = \pi_1(X)$. In case $n = 2$, let $\{\alpha_j\}_{j \in J}$

be a set of generators for the group $\pi_2(\phi)$. If $n \geq 3$ let $\{\alpha_j\}_{j \in J}$ be a set of generators for the module $\pi_n(\tilde{\phi}) \approx \pi_n(\phi)$ over $\mathbb{Z}\pi$. Let

$$\begin{array}{ccc} S^{n-1} & \xrightarrow{f_j} & K \\ \downarrow & & \downarrow \phi \\ E^n & \xrightarrow{g_j} & X \end{array}$$

represent $\alpha_j \in \pi_n(\phi)$. Let $L = K \bigcup_{f_j} (e_j^n)_{j \in J}$. Then there exists an extension $\psi : L \to X$ of ϕ where $\psi | e_j^n$ arises from g_j.

Lemma 4.2. *The map $\psi : L \to X$ is n-connected.*

Proposition 4.3. *Let K be a CW-complex, X of the homotopy type of a CW-complex and $\phi : K \to X$ an $(n-1)$-connected map with $n \geq 2$. Then there exists a CW-complex $\Gamma = K \cup$ (cells of dimension $\geq n$) and a homotopy equivalence $\theta : \Gamma \to X$ extending ϕ.*

$$\begin{array}{ccccc} K & = & K & \stackrel{i}{\hookrightarrow} & L \\ i \downarrow & & \downarrow \phi & & \downarrow \psi \\ L & \xrightarrow{\psi} & X & = & X \end{array}$$

Let $a : (L, K) \to (M_\psi, K)$ and let $b : (M_\phi, K) \to (M_\psi, L)$ be the inclusion maps. Note that the inclusion $M_\phi \hookrightarrow M_\psi$ is a homotopy equivalence since X is a strong deformation retract of both M_ϕ and M_ψ. In particular, this inclusion induces isomorphisms of homotopy groups. From the homotopy exact sequence associated to the triple (M_ψ, M_ϕ, K) we see that $b_* : \pi_k(M_\phi) \to \pi_k(M_\psi)$ is an isomorphism for all k.

Using this isomorphism and the homotopy exact sequence associated to the triple (M_ψ, L, K), we obtain:

$$\cdots \to \pi_k(L, K) \xrightarrow{a_*} \pi_k(\phi) \xrightarrow{b_*} \pi_k(\psi) \xrightarrow{\delta} \pi_{k-1}(L, K) \xrightarrow{a_*} \cdots.$$

Lemma 1.3 follows from this exact sequence. By applying Lemma 1.3 repeatedly we get CW-complexes K_l and maps $\phi_l : K_l \to X$ for $l \geq n$ satisfying

(i) $K_n = K \cup$ (cells of dimension n)
(ii) $K_{l+1} = K_l \cup$ (cells of dimension $l+1$) for $l \geq n$
(iii) $\phi_n | K = \phi$
(iv) $\phi_{l+1} | K_l = \phi_l$ for $l \geq n$ and
(v) ϕ_l is l-connected for $l \geq n$.

If $\Gamma = \bigcup_{l \geq n} K_l$ and $\theta : \Gamma \to X$ is given by $\theta_{|K_l} = \phi_l$ then $\theta : \Gamma \to X$ is homotopy equivalence. That is a consequence of two facts : (1) θ is n-connected for all n. (2) Γ is a CW-complex and X is homotopically equivalent to a CW-complex.

Proposition 4.4. *Let $n \geq 3$, $f : K = K^{n-1} \to X = X^n$ be an $(n-1)$-connected map with K and X finite complexes. Then $\pi_n(f)$ is a finitely generated projective module over $\mathbb{Z}\pi$ where $\pi = \pi_1(X)$.*

Proposition 4.5. *Let $\phi : K = K^2 \to X = X^2$ be a map inducing an isomorphism on fundamental groups. Let K and X be finite complexes. Then $\pi_2(\phi)$ is a finitely generated module over $\mathbb{Z}\pi$ where $\pi = \pi_1(X)$.*

The proofs given by C.T.C.Wall in [11] for Propositions 1.5 and 1.6 are highly geometric and rely on some hard results of J.H.C. Whitehead on simple homotopy types. In Chapter 6, §1 of [10], I have given easy algebraic proofs of these propositions.

Let X be a 0-connected space of the homotopy type of a CW-complex. In what follows π will denote $\pi_1(X)$. We will describe certain conditions on X that will ensure that X is homotopy equivalent to a CW complex K with K^n finite. However, these conditions will be broken into three "groups" depending on whether $n = 1, 2$ or 3.

Definition 4.6.

(i) *We say that X satisfies F_1 if π is finitely generated.*
(ii) *We say that X satisfies F_2 if π is finitely presented and for any finite complex $K = K^2$ and any map $\phi : K^2 \to X$ inducing an isomorphism of fundamental groups, $\pi_2(\phi)$ is a finitely generated over $\mathbb{Z}\pi$.*
(iii) *Let $n \geq 3$. We say that X satisfies F_n if it satisfies F_{n-1} and for any $(n-1)$-connected map $\phi : K = K^{n-1} \to X$ with K finite, $\pi_n(\phi)$ is finitely generated over $\mathbb{Z}\pi$.*

Definition 4.7. *We say that X satisfies F_∞ if X satisfies F_n for all $n \geq 1$.*

The following results were proved by C.T.C.Wall [11].

Theorem 4.8. *Let n be an integer ≥ 1. The following conditions on X are equivalent.*

(i) *X is homotopically equivalent to a CW-complex with finite n-skeleton.*
(ii) *X is dominated by a CW-complex with finite n-skeleton.*
(iii) *X satisfies F_n.*

Theorem 4.9. *The following conditions on X are equivalent.*

(i) *X is homotopically equivalent to a CW-complex of finite type.*
(ii) *X is dominated by a CW-complex of finite type.*
(iii) *X satisfies F.*

Complete proofs of Theorems 4.9 and 4.10 are also given in §1, Chapter 6 of [10].

Remark 4.10.

(a) Given X and any integer $n \geq 2$, there exists an n-connected map $f : K \to X$ with K a CW-complex of dimension $\leq n$.
(b) If $\pi = \pi_1(X)$ is finitely presented, there exists a *finite* complex $K = K^2$ and a map $\phi : K \to X$ with $\phi_* : \pi_1(K) \to \pi_1(X)$ an isomorphism.
(c) Suppose X satisfies F_n with $n \geq 2$. The proof of (iii) $\Rightarrow$ (i) of Theorem 4.9 of the present notes as presented in §1, Chapter 6 of [10] yields an n-connected map $\phi : K = K^n \to X$ with K finite.

(d) Let $n \geq 2$ and X satisfy F_n. Then from (c) above there exists an n-connected map $\phi : K = K^n \to X$ with K finite. Suppose $\pi_{n+1}(\phi)$ is a finitely generated $\mathbb{Z}\pi$-module. Then Milnor's construction yields an $(n+1)$ connected map $\phi_{n+1} : K_{n+1} = K\cup$ (finitely many $(n+1)$ cells)$\to X$. By Proposition 1.4 there exists a CW-complex $\Gamma = K_{n+1}\cup$ (cells of dimension $\geq n+2$) and a homotopy equivalence $\theta : \Gamma \to X$ extending ϕ_{n+1}. Clearly, $\Gamma^{n+1} = K_{n+1}$ is finite. From (i) $\Rightarrow$ (iii) of Theorem 4.9 we see that X satisfies F_{n+1}. Hence for every n-connected map $f : L \to X$ with $L = L^n$ finite we will have $\pi_{n+1}(f)$ finitely generated over $\mathbb{Z}\pi$.

(e) Let π be finitely presented. Then there exists a finite complex $K = K^2$ and a map $\phi : K \to X$ with $\phi_* : \pi_1(K) \to \pi_1(X)$ an isomorphism. If $\pi_2(\phi)$ is finitely generated over $\mathbb{Z}\pi$ we can extend ϕ to a 2-connected map $\phi_2 : K_2 = K\cup$ (finitely many 2-cells) $\to X$. By proposition 1.4, there exists a homotopy equivalence $\theta : \Gamma \to X$ extending ϕ_2 with $\Gamma = K_2\cup$ (cells of dimension ≥ 3). Then $\Gamma^2 = K_2$ is finite. By (i) $\Rightarrow$ (iii) of Theorem 4.9 of the present notes X satisfies F_2. Hence for any $f : L = L^2 \to X$ with L finite and $f_* : \pi_1(L) \to \pi_1(X)$ an isomorphism, we have $\pi_2(f)$ finitely generated over $\mathbb{Z}\pi$.

5. Homotopy type of finite dimensional (resp. finite) complexes

We will deal with conditions for X to be of the homotopy type of a CW-complex K of dimension $\leq n$. When n is an integer ≥ 3, there is a satisfactory answer.

Let X be homotopically equivalent to a complex K of dimension $\leq n$. Then $H_i(\widetilde{X}) = 0$ for $i > n$ and $H^{n+j}(X, \underline{B}) = 0$ for $j \geq 1$ and any local coefficient system $\underline{B}$ over X.

Definition 5.1. *We say that X satisfies D_n if*

(i) $H_i(\widetilde{X}) = 0$ *for* $i > n$ *and*
(ii) $H^{n+1}(X, \underline{B}) = 0$ *for any local coefficient system* $\underline{B}$ *over* X.

Proposition 5.2. *Let $n \geq 3$ and $\phi : K = K^{n-1} \to X$ be an $(n-1)$-connected map (K is not assumed to be finite.) Suppose $H^{n+1}(X, \underline{B}) = 0$ for every local coefficient system $\underline{B}$ over X. Then $\pi_n(\phi)$ is a projective $\mathbb{Z}\pi$-module.*

Proposition 5.3. *Let X satisfy $H_i(\widetilde{X}) = 0$ for $i > n$ where $n \geq 3$. Suppose $\phi : K = K^{n-1} \to X$ is an $(n-1)$-connected map with $\pi_n(\phi)$ a free $\mathbb{Z}\pi$-module. Let $\{\alpha_j\}_{j\in J}$ be a basis for $\pi_n(\phi)$ over $\mathbb{Z}\pi$.*

Let

$$\begin{array}{ccc} S^{n-1} & \xrightarrow{f_j} & K \\ \downarrow & & \downarrow \phi \\ D^n & \xrightarrow[g_j]{} & X \end{array}$$

represent α_j. Let $\psi : K^{n-1}\cup_{f_j}(e^n_j)_{j\in J} \to X$ be the extension of ϕ described in Milnor's construction. Then ψ is a homotopy equivalence.

Lemma 5.4. *Let P be a projective module over a ring R. Then there exists a free R-module F with $P \oplus F \approx F$.*

Theorem 5.5. *Let $n \geq 3$. Then X is homotopically equivalent to a complex L with* $\dim L \leq n$ *if and only if X satisfies D_n.*

Corollary 5.6. *Let $n \geq 3$ and $1 \leq r \leq n-2$. Suppose X satisfies D_n and F_r. Then there exists a homotopy equivalence $\theta : L \to X$ with* $\dim L \leq n$ *and L^r finite.*

Let $n \geq 3$. Assume X satisfies D_n and F_n. Then since X satisfies F_{n-1}, by Remark 1.11 (c) there exists an $(n-1)$ connected map $\phi : K = K^{n-1} \to X$ with K finite. Since X satisfies D_n, from proposition 2.2 we see that $\pi_n(\phi)$ is a projective $\mathbb{Z}\pi$-module. Since X satisfies F_n it follows that $\pi_n(\phi)$ is finitely generated over $\mathbb{Z}\pi$. Thus $P = \pi_n(\phi)$ determines a certain element $[P] \in \widetilde{K}_0(\mathbb{Z}\pi)$.

Choose an epimorphism $\epsilon : F \to P$ in $\mathbb{Z}\pi$-mod with F free on a finite basis $\{X_i\}_{i=1}^k$ so that $\{\epsilon(X_i)\}_{i=1}^k$ generates P over $\mathbb{Z}\pi$. Let

$$\begin{array}{ccc} S^{n-1} & \xrightarrow{f_i} & K \\ \downarrow & & \downarrow \phi \\ E^n & \xrightarrow{g_i} & X \end{array}$$

represent $\alpha_i \in \pi_n(\phi)$. Then the extension $\psi : L = K^{n-1} \cup_{f_i} (e_i)_{i=1}^k \to X$ of ϕ given by Milnor's construction is an n-connected map and $L = L^n$ is a finite complex. Writing the exact homology sequence of $(\widetilde{L}, \widetilde{K})$ we see that $H_{n+1}(\widetilde{\psi}) \oplus P \approx F$ where $\widetilde{\psi} : \widetilde{L} \to \widetilde{X}$ is the lift of ψ to the universal coverings. From the relative Hurewicz Theorem one sees that $\pi_{n+1}(\widetilde{\psi}) \approx H_{n+1}(\widetilde{\psi})$. If we write Q for $\pi_{n+1}(\widetilde{\psi})$, then Q is finitely generated projective over $\mathbb{Z}\pi$ and $P \oplus Q \approx F$. Thus $[\pi_{n+1}(\psi)] = [\pi_{n+1}(\widetilde{\psi})] = -[P] = -[\pi_n(\phi)]$ in $\widetilde{K}_0(\mathbb{Z}\pi)$.

Lemma 5.7. *Let X satisfy F_n and D_n for some $n \geq 3$. As seen above there exists a $\psi : L = L^n \to X$ which is n-connected with L finite. Moreover $\pi_{n+1}(\psi)$ is finitely generated projective over $\mathbb{Z}\pi$. From Remark 1.11 (d) we see that X satisfies F_{n+1}. Also from Theorem 2.5, X is homotopically equivalent to a complex of dimension $\leq n$, since it satisfies D_n. It follows that X satisfies D_{n+1} as well.*

Proposition 5.8. *Let X satisfy F_n and D_n with $n \geq 3$. Let $\psi_i : L_i = L_i^n \to X$ be n-connected with L_i finite for $i = 1, 2$ and $Q_i = \pi_{n+1}(\psi_i)$. Then $[Q_1] = [Q_2]$ in $\widetilde{K}_0(\mathbb{Z}\pi)$.*

Lemma 5.9. *Let X satisfy D_n with $n \geq 3$ and $\psi : L \to X$ an n-connected map. Then there exists an $s : X \to L$ with $\psi \circ s \simeq Id_X$. Hence X is dominated by L.*

In particular, if X satisfies D_n and F_n with $n \geq 3$, then X is dominated by a finite complex.

Theorem 5.10. *X is dominated by a finite complex if and only if X satisfies F_{n_0} and D_{n_0} for some $n_0 \geq 3$. In this case X satisfies F_n and D_n for all $n \geq n_0$. Moreover, for any $n \geq n_0$ there exists an n-connected map $\psi : L = L^n \to X$ with L finite. In this case $\pi_{n+1}(\psi)$ is finitely generated projective over $\mathbb{Z}\pi$. The element*

$\widetilde{\omega}(X) = (-1)^{n+1}[\pi_{n+1}(\psi)] \in \widetilde{K}_0(\mathbb{Z}\pi)$ depends only on the homotopy type of X. The vanishing of $\widetilde{\omega}(X)$ is necessary and sufficient for X to be homotopically equivalent to a finite complex.

Definition 5.11. *When X is dominated by a finite complex, the element $\widetilde{\omega}(X) \in \widetilde{K_0}(\mathbb{Z}\pi)$ is called the Wall obstruction of X.*

Theorem 5.12. *Let $n \geq 3$ and $K = K^{n-1}$ be any finite complex of dimension $\leq n-1$. Let $\pi = \pi_1(K)$ and x any element of $\widetilde{K}_0(\mathbb{Z}\pi)$. Then there exists a finitely dominated CW-complex X with $X^{n-1} = K$ and $\widetilde{\omega}(X) = x$.*

As mentioned already all the results stated in this section were proved by C.T.C. Wall in [11]. Complete proofs can also be found in [10].

6. Integral elements and Dedekind domains

Two good references for the material in this section and the following are [1] and [14]. All the rings we consider in this section onwards will be commutative rings with $1 \neq 0$. When we talk of a sub ring A of B it will be tacitly assumed that $1_A = 1_B$. Usually, the word ideal will mean a proper ideal.

Let A be a sub ring of B and $x \in B$. The polynomial ring in an indeterminate X over A will be denoted by $A[X]$. The sub ring of B generated by A and x will be denoted by $A[x]$. It consists of elements of B which can be written as $a_k x^k + \cdots + a_1 x + a_0$ with $a_i \in A$ for $0 \leq i \leq k$ and k an integer ≥ 0 where x^0 is the element $1 \in B$.

Definition 6.1. *An element $x \in B$ is said to be integral over A if there exists a monic polynomial $f(X) \in A[X]$ with $f(x) = 0$, namely if there exist element $a_0, \ldots, a_{n-1}$ in A satisfying $x^n + a_{n-1}x^{n-1} + \cdots + a_0 = 0$ in B.*

Observe that every element in A is integral over A.

Proposition 6.2. *The following conditions on an element $x \in B$ are equivalent.*

(i) *x is integral over A.*
(ii) *The sub ring $A[x]$ of B is a finitely generated A-module.*
(iii) *There exists a sub ring R of B with $x \in R$ and R finitely generated as an A-module.*
(iv) *There exists a finitely generated A-submodule M of B with $xM \subset M$ and such that $\{\lambda \in A[x] | \lambda M = 0\} = 0$.*

Proposition 6.3. *Let $x_1, \ldots, x_k$ be any finite number of elements of B which are integral over A. Then the sub ring $A[x_1, \ldots, x_k]$ of B is finitely generated as an A-module.*

Corollary 6.4. *The set $\{x \in B | x \text{ integral over } A\}$ is a sub ring of B. It is called the integral closure of A in B and is denoted by $\bar{A}_B$ or $\bar{A}$ if B is clear from the context.*

If $\bar{A} = A$ we say that A is integrally closed in B. If $\bar{A} = B$ then we say that B is integral over A.

Proposition 6.5. *Let $A \subset B \subset C$ be a tower of rings. If B is integral over A and C is integral over B then C is integral over A.*

Definition 6.6. *An integral domain R is said to be integrally closed if it is integrally closed in its quotient field K.*

Definition 6.7. *An integral domain R is called a Dedekind domain if it satisfies the following conditions:*

(i) *R is noetherian*
(ii) *Every prime ideal $P \neq 0$ in R is maximal and*
(iii) *R is integrally closed.*

Every principal ideal domain (abbreviated as PID) is a Dedekind domain.

Definition 6.8. *Any finite extension field K of $\mathbb{Q}$ is called an algebraic number field.*

If A is a sub ring of B, it is clear that the integral closure $\bar{A}_B$ of A in B is integrally closed in B.

Definition 6.9. *Let K be an algebraic number field. An element $\alpha \in K$ is called an algebraic integer in K if α is integral over $\mathbb{Z}$.*

Let R denote the ring of integers in an algebraic number field K. Then R is integrally closed in K. Moreover K is the quotient field of R. The following results are known. Complete proofs can be found in [1], [14] and §2, Chapter 4 of [10]. In these results K is an algebraic number field with $(K : \mathbb{Q}) = n$ and R the ring of integers in K.

Proposition 6.10. *Let $\alpha \in K$ and $Irr(\alpha, \mathbb{Q})$ the irreducible polynomial of α over $\mathbb{Q}$ with leading coefficient 1. Then α is an algebraic integer if and only if $Irr(\alpha, \mathbb{Q}) \in \mathbb{Z}[X]$.*

Theorem 6.11. *R is a free $\mathbb{Z}$-module of rank n. In particular R is noetherian as a $\mathbb{Z}$-module, hence noetherian as a ring.*

Theorem 6.12. *Every non-zero prime ideal of R is a maximal ideal of R.*

Combining these results we see that R is a Dedekind domain.

7. Fractional ideals and the class group

In what follows R will be an integral domain and K will be the quotient field of R. Also R^* (resp. K^*) will denote the set of non-zero elements of R (resp. K).

Definition 7.1. *A fractional ideal of R is a non-zero R-submodule A of K with the property that there exists some $d \in R^*$ (depending on A) with $dA \subseteq R$.*

In this case dA is a non-zero R-submodule of R, hence either $dA = R$ or dA is a non-zero ideal of R. Also, $KA = K$ because there is a non-zero element in A. For any $\lambda \in K^*$, $R\lambda$ is a fractional ideal of R. In fact, if $\lambda = x/y$ with x, y in R^*, we have $yR\lambda \subseteq R$. If $A \subseteq R$, it is called integral. R itself is an integral fractional ideal.

Definition 7.2. *A fractional ideal of R of the from $R\lambda$ with $\lambda \in K^*$ is called a principal fractional ideal.*

Let J denote the set of fractional ideals of R. For A, B in J it is easy to see that $A \cap B$, $A + B$ and AB are fractional ideals.

Definition 7.3. *For any A, B in J we define $(A : B) = \{\lambda \in K \mid \lambda B \subseteq A\}$.*

It is easily checked that $(A : B) \in J$. Also J is a commutative monoid under multiplication given by $A.B = AB$ with R as the identity element.

Definition 7.4. *A fractional ideal A is said to be invertible if there exists a fractional ideal B with $AB = R$.*

Clearly every principal fractional ideal is invertible. For any $\lambda \in K^*$, $R\lambda R\lambda^{-1} = R$. Since J is a commutative monoid, when an element of J is invertible, it has a unique inverse.

Lemma 7.5. *If $A \in J$ is invertible, then $A^{-1} = (R : A)$.*

Lemma 7.6. *If $A \in J$ is invertible, then A is finitely generated as an R-module.*

Lemma 7.7. *Every fractional ideal of R is invertible if and only if every integral ideal of R is invertible. Thus J is a group if and only if every integral ideal of R is invertible.*

Lemma 7.8. *Let R be a noetherian integral domain with K as its quotient field. Then $A \subset K$ is a fractional ideal of R if and only if A is a non-zero finitely generated R submodule of K.*

For the remainder of this section R will be assumed to be a Dedekind domain with quotient field K.

Lemma 7.9. *Let A be an integral ideal of R. Then there exist a finite number of non-zero prime ideals $P_1, \ldots, P_k$ of R with $A \supseteq P_1 P_2 \ldots P_k$.*

Lemma 7.10. *Let A be a non-zero proper ideal of R. Then $(R : A) = \{\lambda \in K \mid \lambda A \subseteq R\}$ satisfies $R \subsetneq (R : A)$.*

Lemma 7.11. *Let A be an integral ideal of R. Then $(R : A)A = R$.*

In particular every integral ideal, hence every fractional ideal of R is invertible. Thus J is a group.

Definition 7.12. *Two fractional ideals A, B of R are said to be equivalent if there exists a $\gamma \in K^*$ with $A = \gamma B$.*

Proposition 7.13. *Two fractional ideals A, B of R are isomorphic in R-mod if and only if they are equivalent in the sense of definition 2.12.*

Lemma 7.14. *Let R be an integral domain (R need not be a Dedekind domain) and I an integral ideal of R. If I is invertible, then I is projective.*

When R is a Dedekind domain, since every integral ideal of R is invertible, from lemma 2.14, we conclude that every ideal of R is projective, hence R is a hereditary ring. From Kaplansky's Theorem, every finitely generated projective module over R is isomorphic to a direct sum of a finite number of ideals in R, some of which could be R.

Let $\mathcal{N} = \{\mathcal{A} \in \mathcal{J} | \mathcal{A} = \mathcal{R}\gamma$ for some $\gamma \in K^*\}$. Then $\mathcal{N}$ is a subgroup of J.

Definition 7.15. *The quotient group $J/\mathcal{N}$ is called the class group of R and is denoted by $Cl(R)$.*

Before stating the main theorem of this section we need the following result.

Theorem 7.16. *Let $M = I_1 \oplus \cdots \oplus I_m$ and $N = J_1 \oplus \cdots \oplus J_n$ where I_i, J_j are fractional ideals of R. Then $M \approx N$ in R-mod if and only if $m = n$ and $\gamma I_1 \ldots I_m = J_1 \ldots J_n$ for some $\gamma \in K^*$.*

For any finitely generated projective R-module $P \approx I_1 \oplus \cdots \oplus I_n$ let $\langle\langle P \rangle\rangle$ denote the class of $I_1 I_2 \ldots I_n$ in $Cl(R)$. If $P \oplus R^k \approx Q \oplus R^l$ with P, Q finitely generated projective R-modules then

$$\langle\langle P \rangle\rangle = \langle\langle P \oplus R^k \rangle\rangle = \langle\langle Q \oplus R^l \rangle\rangle = \langle\langle Q \rangle\rangle.$$

Hence $[P] \rightarrowtail \langle\langle P \rangle\rangle$ is a well-defined homomorphism $\widetilde{K}_0(R) \to Cl(R)$. As an immediate consequence of Theorem 2.16 we get the following.

Theorem 7.17. *For any Dedekind domain R the map $\widetilde{K}_0(R) \to Cl(R)$ is an isomorphism.*

Example 7.18. Let $\omega = e^{2\pi i/p}$ where p is a prime number and $i = \sqrt{-1}$. The ring of integers in the cyolotomic extension $\mathbb{Q}(\omega)$ of $\mathbb{Q}$ is $\mathbb{Z}[\omega]$. For $p = 23$, it is known that $Cl(\mathbb{Z}[\omega]) \neq \{1\}$.

We will accept the following result of Dock Sang Rim [6] without proof.

Theorem 7.19. *Let π be a cyclic group of order p, a prime. Let x denote a fixed generator of π and $\omega = e^{2\pi i/p}$. Then the surjective ring homomorphism $f : \mathbb{Z}\pi \to \mathbb{Z}[\omega]$ carrying x to ω induces an isomorphism $\tilde{K}_0(\mathbb{Z}\pi) \approx \tilde{K}_0(\mathbb{Z}[\omega])$. In particular for $p = 23$, $\tilde{K}_0(\mathbb{Z}\pi) \neq 0$.*

For a proof of the result that $f_* : \tilde{K}_0(\mathbb{Z}\pi) \to \tilde{K}_0(\mathbb{Z}[\omega])$ is an isomorphism, please refer to [6]. A complete proof of this result is also given in §5, Chapter 5 of [10].

Theorem 7.20. *There exists a finitely dominated CW-complex which is not of the homotopy type of any finite CW-complex.*

Proof. Let π be a cyclic group of order 23 and K^2 a finite CW-complex with $\pi_1(K^2) \approx \pi$. From Theorem 2.12, Chapter II of the present notes, there exists a CW-complex X with $X^2 = K$ further satisfying the condition that X is finitely dominated and $\widetilde{\omega}(X) \neq 0$. □

8. Unreduced Wall obstruction

Before embarking on the study of finitely dominated nilpotent spaces we will introduce the "unreduced Wall obstruction" $\omega(X)$ of a finitely dominated space X.

Let R denote a ring. All the chain complexes we consider will be in R-mod and they will be positive chain complexes.

Let $C = \{C_n, d_n\}$ be a chain complex.

Definition 8.1.

(i) *C will be said to be of finite type if each C_n is finitely generated in R-mod.*
(ii) *C will be called a free chain complex if each C_n is a free R-module.*
(iii) *C will be called a finite projective complex (abbreviated as an FP complex) if each C_n is finitely generated projective over R and $C_i = 0$ for $i > n_0$ for some n_0.*
(iv) *C will be called a finite free complex (abbreviated as an FF complex) if each C_n is finitely generated free over R and $C_i = 0$ for $i > n_0$ for some n_0.*

Proposition 8.2. *Let $C = \{C_n, d_n\}$ and $\{C'_n, d'_n\}$ be FP complexes, $f : C \to C'$ a chain map inducing isomorphisms in homology $\{$i.e. $H_*(f) : H_*(C) \approx H_*(C')\}$. Then,*

$$\sum_{i\geq 0}(-1)^i[C_i] = \sum_{i\geq 0}(-1)^i[C'_i] \text{ in } K_0(R).$$

Proposition 8.3. *Let X be dominated by a finite complex and $\pi = \pi_1(X)$. Then the singular chain complex $CS(\widetilde{X})$ of the universal covering $\widetilde{X}$ of X is chain equivalent over $\mathbb{Z}\pi$ to an FP complex $P_* = \{P_n, d_n\}$. The element $\sum_{i\geq 0}(-1)^i[P_i] \in K_0(\mathbb{Z}\pi)$ depends only on the homotopy type of X. Its image in $\widetilde{K}_0(\mathbb{Z}\pi)$ under the canonical quotient map $\eta : K_0(\mathbb{Z}\pi) \to \widetilde{K}_0(\mathbb{Z}\pi)$ is the Wall obstruction $\widetilde{\omega}(X)$ of X (Definition 2.11, Chapter II).*

Definition 8.4. *The element $\omega(X) \in K_0(\mathbb{Z}\pi)$ given by $\omega(X) = \sum_{i\geq 0}(-1)^i[P_i]$ where $P_* = \{P_n, d_n\}$ is any FP complex chain equivalent to $CS(\tilde{X})$ over $\mathbb{Z}\pi$, is referred to as the unreduced Wall obstruction of X. Its image $\eta(\omega(X)) = \widetilde{\omega}(X) \in \widetilde{K}_o(\mathbb{Z}\pi)$ will be referred to as the reduced Wall obstruction of X.*

9. Nilpotent spaces

We first introduce the notion of nilpotent action of a group π on another group G. Let π act on G via $\theta : \pi \to \text{Aut}(G)$.

Definition 9.1. *The lower π-central series of the π-group G is defined as follows : $\Gamma^1_\pi(G) = G$, $\Gamma^2_\pi(G) =$ the smallest normal π-subgroup of G containing $[G, G]$ and $\{\theta(x)(g).g^{-1} | x \in \pi, g \in G\}$. For $k \geq 3$, $\Gamma^k_\pi(G) =$ the smallest normal π-subgroup of G containing $[G, \Gamma^{k-1}_\pi(G)]$ and $\{\theta(x)(g).g^{-1} | x \in \pi, g \in \Gamma^{k-1}_\pi(G)\}$.*

Definition 9.2. *The action of π on G is said to be nilpotent if there exists an integer $k \geq 1$ with $\Gamma^k_\pi(G) = \{1\}$.*

Observe that $\Gamma^k_\pi(G) \supseteq \Gamma^k(G)$. Hence if π acts nilpotently on G, then G itself has to be nilpotent. Thus if G is not nilpotent and even if π acts trivially, this action of π on G is not nilpotent.

Suppose M is a π-module, namely M is an abelian group on which π acts. Then $\Gamma^1_\pi(M) = M$ and $\Gamma^{k+1}_\pi(M) = (I\pi)^k M$ where $I\pi$ is the augmentation ideal in $\mathbb{Z}\pi$. Thus M is a nilpotent π-module if and only if $(I\pi)^k M = 0$ for some integer $k \geq 1$.

Let $G = \pi$ and π act on π by inner automorphisms, then $\Gamma^k_\pi(\pi) = \Gamma^k(\pi)$ for all $k \geq 1$. Hence π acts nilpotently on π under inner automorphisms if and only if π is nilpotent.

Definition 9.3. *Let G be a π-group. We call G a π-perfect group if $\Gamma^2_\pi(G) = G$.*

Definition 9.4. *An arcwise connected space X is said to be nilpotent if the usual action of $\pi_1(X)$ on $\pi_n(X)$ is nilpotent for all $n \geq 1$. This is equivalent to requiring $\pi_1(X)$ to be nilpotent and $\pi_n(X)$ to be a nilpotent $\pi_1(X)$-module.*

Example 9.5.

(1) Any simply connected space is nilpotent.
(2) A space X is said to be simple if $\pi_1(X)$ is abelian and the usual action of $\pi_1(X)$ on $\pi_n(X)$ is trivial for $n \geq 2$. In fact if a group π acts on itself by inner automorphisms, the action is trivial if and only if π is abelian. Thus the usual action of $\pi_1(X)$ on itself is trivial if and only if $\pi_1(X)$ is abelian.
Any simple space is nilpotent. All H-spaces are simple.
(3) Let n be any integer ≥ 1. The real projective space $P^n\mathbb{R}$ is nilpotent if and only if n is odd. The Grassmann space $G_{n,\infty}$ of n planes in $\mathbb{R}^\infty$ is nilpotent if and only if n is odd.
From now onwards all the spaces we consider will be 0-connected and of the homotopy type of a CW-complex. The class of nilpotent spaces among these will be denoted by $\underline{\underline{NH}}$.

Theorem 9.6. *Let $X \in \underline{\underline{NH}}$. Then the following are equivalent for X.*

(i) *$\pi_n(X)$ is finitely generated for all $n \geq 1$.*
(ii) *$H_n(X)$ is finitely generated for all $n \geq 1$.*
(iii) *X is homotopically equivalent to a CW-complex of finite type.*

Definition 9.7. *A space X is said to be quasi-finite if $H_i(X)$ is finitely generated for all i and there exists an integer l with $H_i(X) = 0$ for $i > l$.*

A major result about nilpotent spaces is the following.

Theorem 9.8. *Let $X \in \underline{\underline{NH}}$. Then X is finitely dominated if and only if X is quasi-finite.*

Another nice result on nilpotent spaces is the following:

Theorem 9.9. *Let $X \in \underline{\underline{NH}}$ and finitely dominated with $\pi_1(X)$ infinite. Then the unreduced Wall obstruction $\omega(X)$ of X is 0. In particular $\widetilde{\omega}(X) = 0$, hence X is of the homotopy type of a finite complex.*

Good references for all the results stated above are [3], [4], [5].

10. Finitely dominated nilpotent spaces with a finite fundamental group

In the present section we will state some results on finitely dominated nilpotent spaces with a finite fundamental group. For this purpose we introduce the concept of a special finite projective complex (SFP complex) over $\mathbb{Z}\pi$ where π is any group. $\mathbb{Z}$ will be considered as a right $\mathbb{Z}\pi$-module with trivial π-action. For any two subgroups $G \subset G'$ of π, and $M \in \mathbb{Z}\pi$-mod the canonical quotient map $\mathbb{Z} \otimes_G M \to \mathbb{Z} \otimes_{G'} M$ will be denoted by $\eta^M_{G,G'}$. If C is a chain complex in $\mathbb{Z}\pi$-mod there is an epimorphism $\eta^C_{G,G'} : \mathbb{Z} \otimes_G C \to \mathbb{Z} \otimes_{G'} C$. When there is no possibility of confusion we merely write $\eta_{G,G'}$ for $\eta^M_{G,G'}$ or $\eta^C_{G,G'}$.

Definition 10.1. *Let Γ denote a set of primes. A homomorphism $f : A \to B$ of abelian groups is called a Γ-isomorphism if* ker f *and coker f are torsion groups with no p-torsion for any $p \in \Gamma$.*

Definition 10.2. *An FP complex C over $\mathbb{Z}\pi$ is said to be special if it satisfies the following condition (S).*

(S): For any two subgroups $G \subset G'$ of π with $[G' : G] = k < \infty$, the map

$$(\eta_{G,G})_* : H_*(\mathbb{Z} \otimes_G C) \to H_*(\mathbb{Z} \otimes_{G'} C)$$

is a $\Gamma(k)$-isomorphism. Here $\Gamma(k)$ denotes the set of primes not dividing k.

An FP complex C over $\mathbb{Z}\pi$ satisfying condition (S) is called an SFP complex.

The following result is proved in [3].

Lemma 10.3. *Let $\phi : Y \to X$ be a covering projection of nilpotent spaces with $\phi_*(\pi_1(Y))$ of finite index k in $\pi_1(X)$. Then $\phi_* : H_*(Y : \mathbb{Z}) \to H_*(X : \mathbb{Z})$ is a $\Gamma(k)$-isomorphism.*

Definition 10.4. *A space X is said to be of type FP if $CS(\widetilde{X})$ is chain equivalent to an FP complex over $\mathbb{Z}\pi$ where $\pi = \pi_1(X)$.*

Proposition 10.5. *Suppose X is a nilpotent space of type FP. Then any FP chain complex C chain equivalent to $CS(\widetilde{X})$ over $\mathbb{Z}\pi$ is an SFP chain complex.*

Let $\phi : \pi \to \pi'$ be an epimorphism of groups, H a subgroup of π' and $G = \phi^{-1}(H)$. Using the ring homomorphism $\phi : \mathbb{Z}\pi \to \mathbb{Z}\pi'$ induced by ϕ we regard $\mathbb{Z}\pi'$ as an object in mod $-\mathbb{Z}\pi$. Let $M \in \mathbb{Z}\pi$-mod. Regard $\mathbb{Z}$ as an object of mod-$\mathbb{Z}\pi'$ as well as of mod-$\mathbb{Z}\pi$ with trivial action.

Lemma 10.6. *There exist well-defined maps (as abelian groups)*

$$\psi^H_M : \mathbb{Z} \otimes_G M \to \mathbb{Z} \otimes_H (\mathbb{Z}\pi' \otimes_\pi M)$$

$$\Phi^H_M : \mathbb{Z} \otimes_H (\mathbb{Z}\pi' \otimes_\pi M) \to \mathbb{Z} \otimes_G M$$

satisfying

$$\psi^H_M(n \otimes_G u) = n \otimes_H (e' \otimes_\pi u)$$

$$\Phi^H_M(n \otimes_H (x' \otimes_\pi u)) = n \otimes_G x u$$

for any $n \in \mathbb{Z}, u \in M, x' \in \pi'$ *where* e' *is the identity element of* π' *and* $x \in \pi$ *any element satisfying* $\phi(x) = x'$. *Moreover* ψ_M^H *and* Φ_M^H *are natural in* M *and are isomorphisms, each being the inverse of the other.*

Proposition 10.7. *Let* $\phi : \pi \to \pi'$ *be an epimorphism of groups and* C *an* SFP *complex over* $\mathbb{Z}\pi$. *Then* $\mathbb{Z}\pi' \otimes_\pi C$ *is an* SFP *complex over* $\mathbb{Z}\pi'$.

Lemma 10.8. *Let* π *be any group and* π' *a subgroup of finite index in* π. *If* C *is an* SFP *complex over* $\mathbb{Z}\pi$, *then* C *is an* SFP *complex over* $\mathbb{Z}\pi'$ *as well.*

Proposition 10.9. *Let* π *be a finite nilpotent group and* C *an* SFP *complex over* $\mathbb{Z}\pi$. *Then* $H_*(C)$ *is a nilpotent* π*-module.*

As a consequence one gets the following.

Theorem 10.10. *Let* X *be a space with* $\pi_1(X)$ *finite and nilpotent. Suppose* $CS(\widetilde{X})$ *is chain equivalent to an* SFP *complex over* $\mathbb{Z}\pi_1(X)$. *Then* X *is nilpotent.*

If X is a space of type FP with $\pi_1(X)$ finitely presented, then X is finitely dominated [5]. Combining this with Proposition 3.9 and Theorem 3.10 one obtains the following.

Theorem 10.11. *Let* X *be a space with* $\pi_1(X)$ *finite and nilpotent. Then* X *is finitely dominated and nilpotent if and only if* $CS(\widetilde{X})$ *is chain equivalent to an* SFP *complex.*

For any abelian group π *let* $\overline{\mathbb{Z}\pi}$ *denote the integral closure of* $\mathbb{Z}\pi$ *in* $\mathbb{Q}\pi$ *and* $j : \mathbb{Z}\pi \to \overline{\mathbb{Z}\pi}$ *denote the inclusion. Let* $D(\mathbb{Z}\pi) = \ker j_*$, *where* $j_* : \widetilde{K}_0(\mathbb{Z}\pi) \to \widetilde{K}_0(\overline{\mathbb{Z}\pi})$ *is induced by the inclusion.*

Theorem 10.12. *Let* π *be a finite abelian group and* C *an* SFP *complex over* $\mathbb{Z}\pi$. *Then* $\widetilde{\omega}(C) \in D(\mathbb{Z}\pi)$.

As an immediate consequence one obtains the following.

Theorem 10.13. *Let* X *be a finitely dominated nilpotent space with* $\pi_1(X)$ *finite abelian. Then* $\widetilde{\omega}(X) \in D(\mathbb{Z}\pi)$.

Proofs of results in §3 can be found in [9] and [10].

References

[1] C. W. Curtis and I. Reiner, Representation Theory of Finite Groups and Associative Algebras, Interscience Publishers (1966).
[2] E. Dror, A generalization of the Whitehead theorem, *Lecture Notes in Math.*, **249** Springer (1971) 13–22.
[3] P. Hilton, G. Mislin and J. Roitberg, Localization of Nilpotent Groups and Spaces, *Math. Studies*, **15**, North Holland (1973).
[4] G. Mislin, Wall's obstruction for nilpotent spaces, *Topology*, **14** (1975) 311–317.
[5] G. Mislin, Finitely dominated nilpotent spaces, *Annals of Math.*, **103** (1976) 547–556.
[6] D. S. Rim, Modules over finite groups, *Annals of Mathematics*, **69** (1959) 700–712.
[7] E. H. Spanier, Algebraic Topology, M. Graw-Hill (1966).
[8] J. R. Stallings, Homology and central series of groups, *J. Alg.*, **2** (1965) 170–181.

[9] K. Varadarajan, Finiteness obstruction for nilpotent spaces, *J. Pure Appl. Algebra*, **13** (1978) 137–146.

[10] K. Varadarajan, The Finiteness obstruction of C. T. C. Wall, CMS Series of monographs and advanced texts John Wiley & Sons, Inc. (1989).

[11] C. T. C. Wall, Finiteness conditions for CW-complexes, *Ann. of Math.*, **81** (1965) 56–69.

[12] J. H. C. Whitehead, Combinatorial homotopy I, *Bull. Amer. Math. Soc.*, **55** (1949) 213–245.

[13] J. H. C. Whitehead, Combinatorial homotopy II, *Bull, Amer. Math. Soc.* **55** (1949) 453–496.

[14] O. Zariski and P. Samuel, Commutative Algebra, Vol 1. Van Nostrand Reinhold (1958).

Perspectives in Geometry and Topology No. 11, 2010, pp. 93–99.

Fox Homotopy Groups, Gottlieb Groups, Rhodes Groups, and Generalized Whitehead Products

Peter Wong

Department of Mathematics, Bates College, Lewiston, ME 04240, U.S.A.
e-mail: pwong@bates.edu

Abstract. In this talk, I will describe some recent joint work with Marek Golasiński (Toruń) and Daciberg Gonçalves (São Paulo) on some classical topics in algebraic topology. This project was motivated by the study of Nielsen type fixed point theory for equivariant maps.

Keywords. Torus homotopy groups, Gottlieb groups, Rhodes groups, generalized Whitehead products.

2000 Subject Classification: Primary: 55Q05; Secondary: 55M20

1. Introduction

In classical Nielsen fixed point theory (see [16]), the study of the fixed points of a map $f : X \to X$ is closely related to its induced homomorphism on $\pi_1(X)$. A certain subgroup $J(X) \subset \pi_1(X)$ now known as the Jiang subgroup (or evaluation subgroup or also known as the first Gottlieb group $G_1(X)$) plays an important role in computing the Nielsen number $N(f)$, which is often the minimal number of fixed points in the homotopy class of f. The notion of $J(X)$ was generalized in [27] in the equivariant setting. The G-action, where G is a finite group (for simplicity), is lifted to an action of an extension group $\hat{G}$ such that the following sequence is short exact.

$$1 \to \pi_1(X) \to \hat{G} \to G \to 1 \tag{1.1}$$

Subsequently, M. Golasiński and D. Gonçalves [9] introduced equivariant Gottlieb groups, thereby generalizing the equivariant Jiang subgroup of [27]. However, these are subgroups of π_n and not super groups like $\hat{G}$ in (1.1).

It turns out that $\hat{G}$ coincides with the first Rhodes group $\sigma_1(X, G)$ [19] or the fundamental group of a transformation group. Later, F. Rhodes [20] generalized his construction and introduced higher homotopy groups $\sigma_n(X, G)$ of a transformation group. Furthermore, Rhodes generalized (1.1) by establishing the following short exact sequence.

$$1 \to \tau_n(X) \to \sigma_n(X, G) \to G \to 1 \tag{1.2}$$

where $\tau_n(X) = n$-th torus homotopy group of R. Fox.

This is the write-up of the talk I gave at the International Conference on Geometry and Topology held at the Indian Institute of Technology Bombay, Mumbai, during August 23–26, 2007.

Subsequent work related to $\sigma_n(X, G)$ includes the papers [28,29] of M. Woo and Y. Yoon on evaluation subgroups of σ_1 and σ_n. Now, from (1.2), what are these torus homotopy groups?

In an attempt to give a more geometric interpretation of the classical Whitehead product, R. Fox introduced the so-called torus homotopy groups in 1945 [7] and with full details in 1948 [8]. The idea is to construct a sufficiently large group of homotopy classes of maps so that the classical Whitehead product can be regarded as a commutator in this new group. More precisely, $\tau_n(X)$ is completely determined by $\pi_k(X)$, $1 \leq k \leq n$, and all the Whitehead products.

Throughout, all spaces are well-pointed and compactly generated. Results stated in this talk can be found in [10–12].

2. Fox torus homotopy groups

Let X be a space and $x_0 \in X$. For $n \geq 1$,

$$\tau_n(X, x_0) := \pi_1(X^{T^{n-1}}, \overline{x_0})$$

where $X^{T^{n-1}} = \text{Map}(T^{n-1}, X)$ so that $X^{\mathbb{S}^1}$ denotes the *free loop space* ΛX of X and $\tau_1(X, x_0) = \pi_1(X, x_0)$. For connected X, τ_n is independent of the basepoint x_0. Let $\pi_i = \pi_i(X)$. The following result is due to R. Fox [8].

Theorem 2.1.

$$0 \to \prod_{i=2}^{n} (\pi_i)^{\alpha_i} \to \tau_n(X) \overset{\leftarrow\,-\,-}{\to} \tau_{n-1}(X) \to 1 \tag{2.1}$$

is split exact where $\alpha_i = \binom{n-2}{i-2}$.

To re-interpret Fox's result, we begin with $n = 2$. It follows that

$$\begin{aligned}\pi_1(X^{\mathbb{S}^1}; \overline{x_0}) &= [\mathbb{S}^1, X^{\mathbb{S}^1}] \\ &\cong [\mathbb{S}^1, \text{Map}_*(\mathbb{S}^1 \sqcup *, X)] \\ &\cong [\mathbb{S}^1 \wedge (\mathbb{S}^1 \sqcup *), X)] \\ &\cong [\Sigma(\mathbb{S}^1 \sqcup *), X].\end{aligned}$$

Note here that $\Sigma(\mathbb{S}^1 \sqcup *)$ is a pinched 2-torus. This suggests that we define

$$\begin{aligned}F_n &:= \Sigma(T^{n-1} \sqcup *) \\ &\approx T^n / T^{n-1}\end{aligned} \tag{2.2}$$

We proved in [10] that the (reduced) suspension of F_n has the homotopy type of a wedge of spheres. More precisely, we have the following result.

Proposition 2.2.

$$\Sigma F_{n-1} \approx \bigvee_{i=2}^{n} (\mathbb{S}^i)^{\binom{n-2}{i-2}}.$$

We call F_n the n-th Fox space (with respect to $\mathbb{S}^1$). Now, we re-define

$$\tau_n(X, x_0) := [\Sigma(T^{n-1} \sqcup *), X].$$

In [10], we established

Theorem 2.3.

$$0 \to \tau_{n-1}(\Omega X) \to \tau_n(X) \overset{\leftarrow\,-\,-}{\to} \tau_{n-1}(X) \to 1$$

is split exact.

We also showed that

$$\tau_{n-1}(\Omega X) \cong \prod_{i=2}^{n} (\pi_i(X))^{\alpha_i}$$

so that we recover Fox's Theorem 2.1. Here are some nice properties of τ_n:

With respect to a fibration $F \hookrightarrow E \to B$, we have the following

Theorem 2.4. *For any positive integer k, there exists a long exact sequence*

$$\cdots \to \tau_k(\Omega^n F) \to \tau_k(\Omega^n E) \to \tau_k(\Omega^n B) \overset{d_*}{\to} \tau_k(\Omega^{n-1} F) \to \cdots$$

For any X, the Fox split exact sequence (2.1) implies that we have a filtration of torus homotopy groups:

$$\tau_1(X) \subseteq \tau_2(X) \subseteq \tau_3(X) \subseteq \ldots$$

It has been shown in [8] that if $\alpha \in \pi_m$, $\beta \in \pi_n$ and $(\alpha, \beta) \in \pi_{m+n-1}$ denotes the Whitehead product of α and β then the image of (α, β) in τ_{m+n-1} is a commutator.

3. Generalization

Using (2.2), we can define the generalized Fox homotopy groups of X with respect to a space W as follows. Let W be a space, $x_0 \in X$. Define

$$\tau_W(X, x_0) := [\Sigma(W \sqcup *), X]$$

When $W = T^{n-1}$, $\tau_W(X) = \tau_n(X)$, the n-th Fox group. Now Theorem 2.3 can be further generalized [10].

Theorem 3.1. *Let X, V, W be path connected spaces. Then,*

$$1 \to [(V \times W)/V, \Omega X] \to \tau_{V\times W}(X) \overset{\leftarrow\,-\,-}{\to} \tau_V(X) \to 1$$

is split exact. If $W = \Sigma A$, then $[(V \times W)/V, \Omega X]$ is abelian and is isomorphic to $[V \wedge W, \Omega X] \times [W, \Omega X]$.

When $W = \mathbb{S}^1$ and $V = T^{n-2}$, $(V \times W)/V \approx \Sigma(T^{n-2} \sqcup *)$. Thus, Theorem 3.1 becomes Theorem 2.3. Furthermore, Fox's interpretation of the Whitehead product as a commutator remains valid in this general setting.

Given $f : \Sigma A \to X$ and $g : \Sigma B \to X$, let

$$f' = f \circ \Sigma p_A \quad \text{and} \quad g' = g \circ \Sigma p_B$$

where

$$\Sigma p_A : \Sigma(A \times B) \to \Sigma A, \; \Sigma p_B : \Sigma(A \times B) \to \Sigma B$$

are the suspensions of the corresponding projections. Using the co-multiplication of $\Sigma(A \times B)$, the map

$$K = (f' \cdot g') \cdot (f'^{-1} \cdot g'^{-1}) : \Sigma(A \times B) \to X$$

is well-defined and is homotopic to a constant when restricted to $\Sigma(A \vee B)$. Thus, K induces $K' : \Sigma(A \wedge B) \to X$ and

$$[f] \circ [g] := [K']$$

is the generalized Whitehead product given by M. Arkowitz [2]. (Here, we use the inverses instead.)

Theorem 3.2. *The image of* $[f] \circ [g]$ *in* $\tau_{A \times B}(X)$ *is the commutator of the image of* $[f]$ *and the image of* $[g]$ *in* $\tau_{A \times B}(X)$.

4. Gottlieb groups

In the classical case, where $G_n(X)$ denotes the n-th Gottlieb group of X, the Whitehead products of $\alpha \in G_n(X)$ with any elements in the higher homotopy groups vanish.

Given the group $[\Sigma A, X]$, the *generalized* Gottlieb group is given by

$$\mathcal{G}(\Sigma A, X) = \mathrm{Im}(ev_* : [\Sigma A, (X^X, 1_X)] \to [\Sigma A, (X, x_0)]).$$

It can also be defined as the set of $\alpha : \Sigma A \to X$ such that the map $\alpha \vee 1_X$ can be extended to $\Sigma A \times X$ up to homotopy. It is known that $\mathcal{G}(\Sigma A, X)$ is central in $[\Sigma A, X]$. In fact, K. Varadarajan [26] showed that for all $\alpha \in \mathcal{G}(\Sigma A, X)$ and for all $\beta \in [\Sigma B, X]$

$$\alpha \circ \beta (\in [\Sigma(A \times B), X]) = 0.$$

Using the generalized Fox homotopy groups, we obtain the following:

Theorem 4.1. *The group* $\mathcal{G}(\Sigma A, X)$ *when regarded as a subgroup of* $\tau_{A \times B}(X)$ *is central in* $\tau_{A \times B}(X)$ *for any* B.

Remark 4.1. Although $\tau_k(X)$ is not abelian in general, the evaluation subgroup (Gottlieb-Fox group) $G\tau_n(X) = \mathcal{G}(\Sigma T^{n-1}, X)$ turns out to be abelian and is central in $\tau_n(X)$. In fact,

$$G\tau_n(X) = \prod_{i=1}^{n} (G_i(X))^{\gamma_i}$$

where $\gamma_i = \binom{n-1}{i-1}$ and $G_i(X) = i$-th Gottlieb group.

5. Rhodes groups

We now turn to the higher homotopy groups of a transformation group first introduced by F. Rhodes [20]. Let G be a finite group, X a finite G-complex, and $x_0 \in X$. Denote by $C_n = \{(t_1, \hat{t}_2, \ldots, \hat{t}_n)\}$, the n-dimensional cylinder where $0 \leq t_i \leq 1$ and $\hat{t}_i = t_i$ (mod 1).

A map $f : C_n \to X$ is of *order* g, $g \in G$ if

$$f(0, \hat{t}_2, \ldots, \hat{t}_n) = x_0, \quad f(1, \hat{t}_2, \ldots, \hat{t}_n) = g(x_0).$$

Then, we let $\sigma_n(X, x_0, G)$ be the set of homotopy classes of maps of order g. Together with the operation given by

$$[f_1; g_1] * [f_2; g_2] := [f_1 + g_1 f_2; g_1 g_2],$$

$\sigma_n(X, G)$ becomes a group. The subgroup consisting of elements of order e, the identity of G, turns out to be the torus homotopy group. This fact was proved in [20] and is stated below.

Theorem 5.1.

$$1 \to \tau_k(X, x_0) \to \sigma_k(X, x_0, G) \to G \to 1$$

is exact.

There is another relation between the Rhodes groups and the Fox torus homotopy groups as follows [11].

Theorem 5.2. *If G acts freely on X, then the projection map $p : X \to X/G$ induces an isomorphism*

$$\sigma_k(X, x_0, G) \cong \tau_k(X/G, p(x_0)).$$

Similar to the Gottlieb-Fox group $G\tau_n(X)$, one can define the equivariant Gottlieb groups or Gottlieb-Rhodes groups to be

$$\mathcal{G}_n = \mathrm{Im}(ev_* : \sigma_n(X^X, 1_X, G) \to \sigma_n(X, x_0, G)).$$

M. Woo and Y. Yoon [28] asked whether $\mathcal{G}_n$ is always abelian. In [11], we showed that $\mathcal{G}_n$ is in fact an extension of the Gottlieb-Fox group $G\tau_n$ and we have the following

Theorem 5.3. *Let G_0 be the subgroup of G consisting of elements g which lie in the same component of 1_X and $\mathcal{W}_n = G\tau_n(X)$. Then,*

$$1 \to \mathcal{W}_n \to \mathcal{G}_n \to G_0 \to 1$$

is exact.

Thus, $\mathcal{G}_n$ is not always abelian.

6. Jacobi identities

In any group, there is a well-known Jacobi identity:

$$(b^{-1}, a, c)^b (c^{-1}, b, a)^c (a^{-1}, c, b)^a = 1,$$

where $(x, y) = xyx^{-1}y^{-1}$ and $(x, y, z) = ((x, y), z)$.

For the classical Whitehead product, similar Jacobi identities have been obtained by M. Nakaoka and H. Toda [17], H. Suzuki [23], G. Whitehead [25], S. C. Chang [5], P. Hilton [14], H. Uehara and W. Massey [24]. In [3], W. Barcus and M. Barratt obtained one such identity with one of the elements from π_1. Later, P. Hilton [15] found such an identity in which elements from π_1 are allowed.

In a more general setting, similar identities for the generalized Whitehead product have been obtained by D. Cohen [6], M. Arkowitz [2], J. Boardman and B. Steer [4], H. Ando [1], and the most general form was obtained by J. W. Rutter [21,22].

Using the generalized Fox homotopy groups, we gave the following unified approach to the Jacobi identities of Rutter's [21,22] and of Hilton's [15].

Theorem 6.1. *Let A, B and C be pointed spaces with the homotopy type of compactly generated Hausdorff spaces. Suppose $\alpha \in [\Sigma A, X]$, $\beta \in [\Sigma B, X]$ and $\gamma \in [\Sigma C, X]$ and denote by $\bar{\alpha}, \bar{\beta}$, and $\bar{\gamma}$ the respective images in $\tau_{A\times B\times C}(X)$. Denote by $[-,-]$ the generalized Whitehead product in $\tau_{A\times B\times C}(X)$. Then*

(1) $(\Sigma t_{213})^*[[\bar{\beta}^{-1}, \bar{\alpha}], \bar{\gamma}]^{\bar{\beta}} + (\Sigma t_{321})^*[[\bar{\gamma}^{-1}, \bar{\beta}], \bar{\alpha}]^{\bar{\gamma}} + (\Sigma t_{132})^*[[\bar{\alpha}^{-1}, \bar{\gamma}], \bar{\beta}]^{\bar{\alpha}} = 1$, *where t_{ijk} is the appropriate twisting function for $i, j, k = 1, 2, 3$.*

(2) *If ΣA, ΣB and ΣC are homotopy co-commutative co-H-spaces, then* $[[\alpha, \beta], \gamma] + (\Sigma t_{312})^*[[\gamma, \alpha], \beta] + (\Sigma t_{231})^*[[\beta, \gamma], \alpha] = 0$.

(3) *If each of ΣA, ΣB, and ΣC is either homotopy co-commutative or equal to $\mathbb{S}^1$ then* $(\Sigma t_{213})^*\beta \cdot [[\beta^{-1}, \alpha], \gamma] + (\Sigma t_{321})^*\gamma \cdot [[\gamma^{-1}, \beta], \alpha] + (\Sigma t_{132})^*\alpha \cdot [[\alpha^{-1}, \gamma], \beta] = 1$ *where*

$$\xi \cdot \eta = \begin{cases} \eta, & \text{if } \xi \in [\Sigma W, X] \text{ and } W \neq \mathbb{S}^0; \\ \eta^{\xi}, & \text{if } \xi \in \pi_1(X). \end{cases}$$

7. Concluding remarks

Theorem 6.1 can further be extended to Jacobi identities for Γ-Whitehead products where Γ is any co-grouplike space replacing the unit circle $\mathbb{S}^1$. This approach generalizes the Jacobi identity obtained by N. Oda [18].

For further work on (equivariant) Gottlieb-Rhodes groups, see [13].

References

[1] H. Ando, On the generalized Whitehead products and the generalized Hopf invariant of a composition element, *Tôhoku Math. J. (2)*, **20** (1968) 516–553.

[2] M. Arkowitz, The generalized Whitehead product, *Pacific J. Math.*, **12** (1962) 7–23.

[3] W. Barcus and M. Barratt, On the homotopy classification of the extensions of a fixed map, *Trans. Amer. Math. Soc.*, **88** (1958) 57–74.

[4] J. Boardman and B. Steer, On Hopf invariants, *Comment. Math. Helv.*, **42** (1967) 180–221.

[5] S. Chang, On Jacobi identity, (Chinese) *Acta. Math. Sinica*, **4** (1954) 365–379.

[6] D. E. Cohen, Products and carrier theory, *Proc. London Math. Soc.*, **7** (1957) 219–248.

[7] R. Fox, Torus homotopy groups, *Proc. Nat. Acad. Sci. (U.S.A.)*, **31** (1945) 71–74.

[8] R. Fox, Homotopy groups and torus homotopy groups, *Ann. Math.*, **49** (1948) 471–510.

[9] M. Golasiński and D. Gonçalves, Equivariant Gottlieb groups, *Cahiers Topologie Géom. Différentielle Catég.*, **42** (2001) 83–100.

[10] M. Golasiński, D. L. Gonçalves and P. Wong, Generalizations of Fox homotopy groups, Whitehead products, and Gottlieb groups, *Ukrain. Math. Zh.*, **57** (2005) no. 3, 320–328 (translated in Ukrainian Math J. **57** (2005) no. 3, 382–393).

[11] M. Golasiński, D. Gonçalves and P. Wong, Equivariant evaluation subgroups and Rhodes groups, *Cahiers Topologie Géom. Différentielle Catég.*, **48** (2007) 55–69.

[12] M. Golasiński, D. Gonçalves and P. Wong, On Fox spaces and Jacobi identities, *Math. J. Okayama U.*, **50** (2008) 161–176.

[13] M. Golasiński, D. Gonçalves and P. Wong, A note on generalized equivariant homotopy groups, Banach Center Publ. **85** (2009) 179–185.

[14] P. Hilton, On homotopy groups of union of spheres, *J. London Math. Soc.*, **30** (1955) 154–172.

[15] P. Hilton, Note on the Jacobi identity for Whitehead products, *Proc. Cambridge Philos. Soc.*, **57** (1961) 180–182.

[16] Boju Jiang, Lectures on Nielsen Fixed Point Theory, *Contemp. Math.*, **14** Amer. Math. Soc., Providence (1983).

[17] M. Nakaoka and H. Toda, On Jacobi identity for Whitehead products, *J. Inst. Polytech. Osaka City Univ. Ser. A*, **5** (1954) 1–13.

[18] N. Oda, A generalization of the Whitehead product, *Math. J. Okayama Univ.*, **39** (1997) 113–133 (1999).

[19] F. Rhodes, On the fundamental group of a transformation group, *Proc. London Math. Soc.*, **16** (1966) 635–650.

[20] F. Rhodes, Homotopy groups of transformation groups, *Canad. J. Math.*, **21** (1969) 1123–1136.

[21] J. W. Rutter, Two theorems on Whitehead products, *J. London Math. Soc.*, **43** (1968) 509–512.

[22] J. W. Rutter, Correction to Two theorems on Whitehead products, *J. London Math. Soc. (2)*, **1** (1969) 20.

[23] H. Suzuki, A product in homotopy theory, *Tôhoku Math. J. (2)*, **6** (1954) 78–88.

[24] H. Uehara and W. Massey, The Jacobi identity for Whitehead products, *Algebraic geometry and topology* (A symposium in honor of S. Lefschetz) pp. 361–377, Princeton University Press, Princeton, N.J. (1957).

[25] G. Whitehead, On mappings into group-like spaces, *Comment. Math. Helv.*, **28** (1954) 320–328.

[26] K. Varadarajan, Generalised Gottlieb groups, *J. Indian Math. Soc.*, **33** (1969) 141–164.

[27] P. Wong, Equivariant Nielsen numbers, *Pacific J. Math.*, **159** (1993) 153–175.

[28] M. Woo and Y. Yoon, Certain subgroups of homotopy groups of a transformation group, *J. Korean Math. Soc.*, **20** (1983) 223–233.

[29] M. Woo, and Y. Yoon, A certain subgroup of the fundamental group of a transformation group, *Bull. Korean Math. Soc.*, **30** (1993) no. 1, 53–59.

Perspectives in Geometry and Topology No. 11, 2010, pp. 101–132.

Combinatorial and Geometric Group Theoretic Methods in Fixed Point Theory

Peter Wong

Department of Mathematics, Bates College, Lewiston, ME 04240, USA
e-mail: pwong@bates.edu

Abstract. During the first half of the twentieth century, the foundations of algebraic topology were laid and the subject dominated the development of modern mathematics. Many basic notions and tools of algebraic topology were developed, partially due to the need for new techniques to solve problems in fixed point theory. In the early 1920's, S. Lefschetz established the field of topological fixed point theory by proving his celebrated Fixed Point Theorem. Subsequently, J. Nielsen, H. Hopf, K. Reidemeister, F. Wecken, W. Franz and others further advanced the subject. Nearly a century later, topological fixed point theory continues to play an important role in applications to dynamics and non-linear analysis, among others. Furthermore, other branches of mathematics have found connections with topological fixed point theory.

These lectures are aimed at presenting some of the techniques from combinatorial and geometric group theory recently used in the study of topological fixed point theory. Most of the results in this paper are stated without proofs and details can be found elsewhere. The paper is organized as follows. First, we give a brief introduction and overview of Nielsen fixed point theory, which provides the motivation for the use of group theoretic techniques. For more details on Nielsen fixed point theory, basic references are [5], [40], and [44] (see also [6], [9], [35]). A brief survey of existing techniques is given. In particular, we emphasize the Reidemeister trace and its computation in Sections 2 and 3. Certain methods from combinatorial group theory and from geometric group theory will then be presented in Sections 4 and 5. These methods are used in determining the infiniteness of the Reidemeister number.

Keywords. Fixed point theory, Nielsen number, Reidemeister number, twisted conjugacy classes, Reidemeister trace, R_∞ property.

Subject Classification: 55M20, 20E45

1. Nielsen fixed point theory

In 1912, L.E.J. Brouwer proved that every continuous mapping from the closed unit n-disk to itself must have a fixed point. Substantially generalizing this result, S. Lefschetz proved in 1921 his celebrated fixed point theorem for manifolds.

This paper is a revised version of a series of five lectures I gave at the International Workshop on Geometry and Topology held at the Indian Institute of Technology Bombay, Mumbai, during August 17–22, 2007. I wish to thank the scientific and organizing committees of the Workshop, in particular, Professor P. Sankaran and Professor A. Shastri, for their kind invitation and to the Indian Institute of Technology Bombay for its hospitality and support during my visit to the Institute, August 14–27, 2007. I thank Peter Zvengrowski for many helpful comments and suggestions.

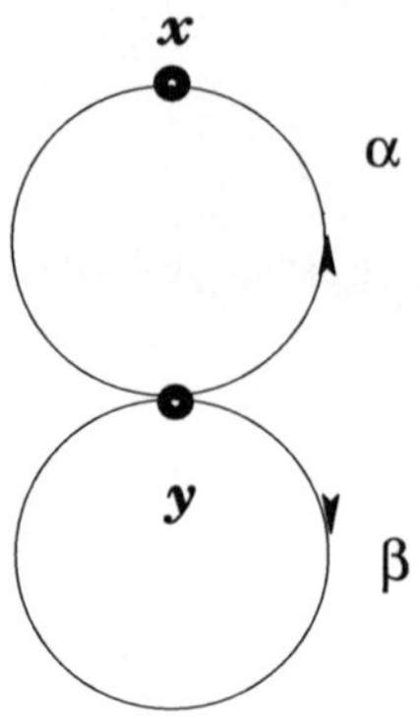

Figure 1. Figure eight

Theorem 1.1 (Lefschetz Fixed Point Theorem). *Let $f : M \to M$ be a self-map of a closed connected oriented triangulable n-manifold M. Define the Lefschetz number to be*

$$L(f) := \sum_{q=0}^{n} (-1)^q tr(f_{*_q} : H_q(M; \mathbb{Q}) \to H_q(M; \mathbb{Q})).$$

If $L(f) \neq 0$ then $Fix(f) := \{x \in M \,|\, f(x) = x\} \neq \emptyset$.

Remark 1.1. Since the Lefschetz number is defined in homological terms, it is invariant under homotopy, that is, $L(f) = L(f')$ for any $f' \simeq f$. Theorem 1.1 holds for more general spaces, e.g., for M an n-dimensional finite simplicial complex.

Unfortunately, the converse of Theorem 1.1 is not true in general.

Example 1.1. Consider the figure eight space X (see Figure 1), a wedge of two circles, based at y. Its fundamental group is free on α, β, the loops representing α, β being shown in Figure 1.

Let $f : X \to X$ be a map such that on the fundamental group, $f_\#(\alpha) = \alpha^{-1}$ and $f_\#(\beta) = \beta^2$ and $Fix(f) = \{x, y\}$. At $H_1(X; \mathbb{Q})$, we have

$$f_{*_1} = \begin{pmatrix} -1 & 0 \\ 0 & 2 \end{pmatrix}.$$

Thus,

$$L(f) = 1 - tr \begin{pmatrix} -1 & 0 \\ 0 & 2 \end{pmatrix} = 1 - 1 = 0.$$

As it turns out, $Fix(h) \neq \emptyset$ for all $h \simeq f$ because its Nielsen number is non-zero. This will be shown in Section 3.

In order to detect *essential* fixed points, we use a more subtle invariant, namely the Nielsen number.

Suppose $f : X \to X$ is a map of a compact connected polyhedron X and $Fix(f) \neq \emptyset$. Then for $x, y \in Fix(f)$, we say that x and y are *Nielsen equivalent as fixed points with respect to f*, if there exists a path $C : [0, 1] \to X$, $C(0) = x$, $C(1) = y$ such that $f \circ C \simeq C$ relative to the endpoints.

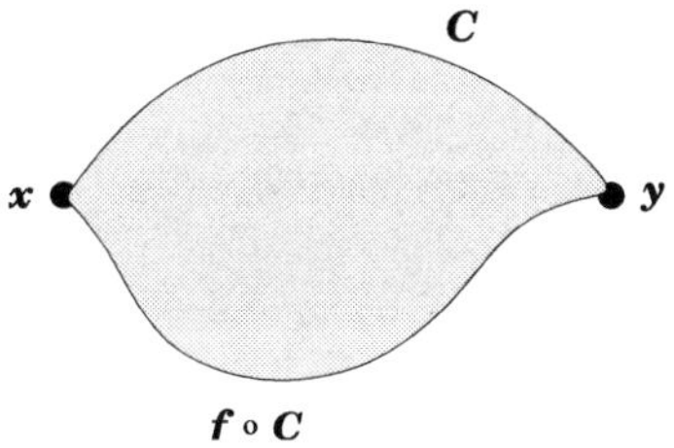

Figure 2. Nielsen equivalence

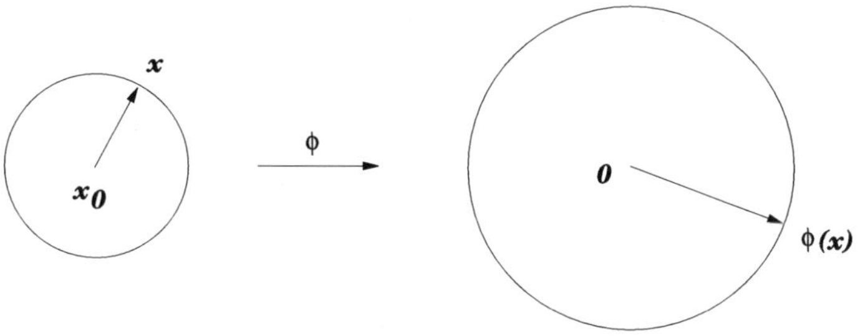

Figure 3. Local fixed point index

This defines an equivalence relation on the set $Fix(f)$. The equivalence classes are called *Nielsen classes* or *fixed point classes* of f. Since X is compact, the Nielsen classes are both open and closed in $Fix(f)$. To detect *essentiality*, we use the usual notion of a fixed point index.

For our purposes, we can make use of the local degree as follows. First of all, we may assume that f has isolated fixed points and hence $\#Fix(f) < \infty$ (e.g. if X is a smooth manifold and f is smooth, we may use transversality to deform f to f' such that the graph of $1 \times f'$ intersects the diagonal ΔX in $X \times X$ transversely), up to homotopy.

Let $x_0 \in Fix(f)$ be a fixed point and we may assume that $X \subset \mathbb{R}^N$ for some N. Choose a sufficiently small euclidean neighborhood $B_r(x_0)$ such that $(Fix(f) - \{x_0\}) \cap B_r(x_0) = \emptyset$ and $B_r(x_0) \cong \mathbb{R}^N$. Moreover, $Fix(f) \cap \partial B_r(x_0) = \emptyset$ so that $x - f(x) \neq 0$ for $x \in \partial B_r(x_0)$. Now let $\phi : \partial B_r(x_0) \to \partial B_1(0)$ be given by

$$\phi(x) = \frac{x - f(x)}{\|x - f(x)\|}$$

for $x \in \partial B_r(x_0)$. Note that $\partial B_r(x_0) \cong \partial B_1(0) \cong S^{N-1}$ carrying the same orientation as that for S^{N-1} so that $\deg \phi : S^{N-1} \to S^{N-1}$ is well-defined. We define the fixed point index of the point x_0 to be $index(f; x_0) := \deg \phi$.

If $\mathbb{F}$ is a fixed point class, then

$$index(\mathbb{F}) := \sum_{x \in \mathbb{F}} index(f; x).$$

For more general spaces than manifolds, a more general fixed point index must be used (see e.g. [5], [35], [40]).

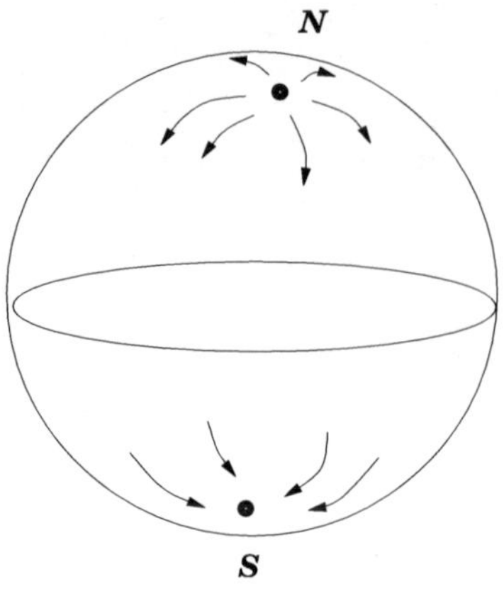

Figure 4. Identity map on S^2

Definition 1.1. *The Nielsen number $N(f)$ of f is defined as the number of essential fixed point classes, that is, the number of fixed point classes $\mathbb{F}$ with index$(\mathbb{F}) \neq 0$.*

Among many properties of the fixed point index, the following two are crucial for our discussions.

(I) (Homotopy Invariance) If $f \simeq f'$ then there is a 1-1 correspondence between the fixed point classes of f and f' as well as their indices.

(II) (Removability) If $index(f; x_0) = 0$ then x_0 can be removed locally, that is, there exists a neighborhood U of x_0 with $U \cap Fix(f) = \{x_0\}$ and a homotopy $f' \simeq_H f$ (relative to the complement of U) such that $H_t|_{X-U} = f$, and $Fix(f') \cap U = \emptyset$.

Remark 1.2. Since $Fix(f)$ is finite, we have

$$L(f) = \sum_{x \in Fix(f)} index(f; x).$$

The Nielsen number also enjoys many properties similar to the Lefschetz fixed point index. Two of these are:

(I) (Lower Bound) $0 \leq N(f) \leq \#Fix(f)$

(II) (Homotopy Invariance) If $f \simeq h$ then $N(f) = N(h)$.

Example 1.2. Take X to be the 2-sphere and let f be the identity map. Then $L(id) = \chi(X) = 2$. Deform the identity to a map f that has the North Pole and the South Pole as fixed points each of which is of fixed point index $+1$ (see Figure 4).

One can further deform f to a map f' with only *one* fixed point, say the North Pole (see Figure 5). Note that N and S are in the same Nielsen class. One can coalesce N and S into one point with the same *total* index, i.e., $index(f'; N) = 2$.

In this simple example, we find a map f' in the homotopy class of $f (\simeq id_{S^2})$ with one fixed point *class* in which there is only *one* fixed point. Furthermore, the

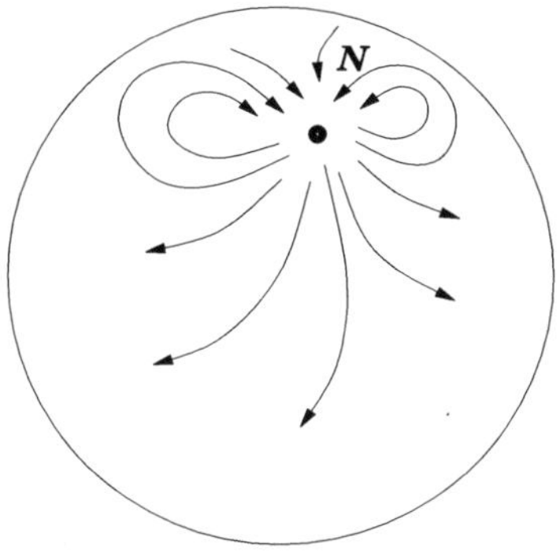

Figure 5. One fixed point of index 2

class is *essential* since the index is non-zero so we cannot remove the point under any homotopy. This is the minimality result.

Given $f : X \to X$, let

$$MF[f] := \min\{\#Fix(h) | h \simeq f\}.$$

By definition, $N(f) \leq MF[f]$. In our example, $N(id_{S^2}) = MF[id_{S^2}] = 1$.

For surface automorphisms, J. Nielsen (1927) asked whether $N(f) = MF[f]$. He also asked whether the equality holds for all surface maps.

In the early 1940's, F. Wecken [63] proved that for many polyhedra of dimension at least 3, $N(f) = MF[f]$. In particular, we have

Theorem 1.2 (Wecken). *Let X be a compact triangulable manifold of dimension $n \geq 3$. For any self-map $f : X \to X$, $N(f) = MF[f]$, that is, there exists $f' \simeq f$ such that $\#Fix(f') = N(f') = N(f)$.*

This means that the Nielsen number is often the best (sharp) lower bound for the minimal number of fixed points.

Remark 1.3. The Nielsen's conjecture for self homeomorphisms holds, that is, for surface automorphisms f, $f \simeq f'$ (isotopy) such that $\#Fix(f') = N(f)$. But for general maps on surfaces of negative Euler characteristic, there exist maps f_n so that $MF[f_n] - N(f_n) > n$ for any positive integer n (see e.g. [6] or [9]).

Question. How do we compute $N(f)$?

For this, we look to the covering space approach of fixed point classes.

In 1927, J. Nielsen gave another description of the fixed point classes via covering space theory as follows. Let $\eta : \tilde{X} \to X$ be the universal cover of a connected compact polyhedron X. The group of deck transformations $Cov(\eta)$ can be identified with the fundamental group $\pi_1(X)$. Choosing base-points and fixing a lift $\tilde{f}$ of a self-map f, every lift of f to the universal cover is of the form $\alpha\tilde{f}, \alpha \in Cov(\eta)$. It is clear that $\eta(Fix(\tilde{f})) \subset Fix(f)$. Furthermore, every $x \in Fix(f)$ is the projection of a fixed point of some lift $\alpha\tilde{f}$. Thus, $Fix(f) = \bigcup_{\alpha \in Cov(\eta)} \eta(Fix(\alpha\tilde{f}))$. In fact, for any two lifts $\tilde{f}_1, \tilde{f}_2$ of f, either $\eta(Fix(\tilde{f}_1)) \cap \eta(Fix(\tilde{f}_2)) = \emptyset$ or $\eta(Fix(\tilde{f}_1)) = \eta(Fix(\tilde{f}_2))$.

Two lifts $\tilde{f}_1, \tilde{f}_2$ yield the same set $\eta(Fix(\tilde{f}_1)) = \eta(Fix(\tilde{f}_2))$ iff $\tilde{f}_2 = \gamma \tilde{f}_1 \gamma^{-1}$ for some $\gamma \in Cov(\eta)$. It is not hard to show that the non-empty sets $\{\eta(Fix(\alpha \tilde{f}))\}$ are precisely the fixed point classes.

This approach was carried further by K. Reidemeister in the 1930's. By the uniqueness of liftings, for any $\alpha \in Cov(\eta)$, there exists a unique element $\varphi(\alpha) \in Cov(\eta)$ such that $\varphi(\alpha)\tilde{f} = \tilde{f}\alpha$. Now suppose $\beta \tilde{f}$ and $\alpha \tilde{f}$ are equivalent, i.e., $\beta \tilde{f} = \gamma \alpha \tilde{f} \gamma^{-1}$. Then

$$\beta \tilde{f} = \gamma \alpha \varphi(\gamma)^{-1} \tilde{f}$$

which implies that $\beta = \gamma \alpha \varphi(\gamma)^{-1}$. Note that $\varphi : Cov(\eta) \to Cov(\eta)$ is a group homomorphism (essentially the same as the induced homomorphism $f_\#$). Then there is a 1-1 correspondence between $\{\eta(Fix(\alpha \tilde{f}))\}/\sim$ and the orbits of the Reidemeister action $\sigma \cdot \alpha \mapsto \sigma \alpha \varphi(\sigma)^{-1}$. From now on, we make the following identification $\pi \equiv Cov(\eta) \equiv \pi_1(X)$.

Let $\mathcal{R}(\varphi)$ be the set of orbits of the action $\sigma \cdot \alpha \mapsto \sigma \alpha \varphi(\sigma)^{-1}$ and $R(\varphi)$ be the cardinality of the set $\mathcal{R}(\varphi)$. Since each φ-twisted conjugacy class corresponds to some $\eta(Fix(\alpha \tilde{f}))$ which may be empty, it follows that $N(f) \le R(f) := R(f_\#)$.

Example 1.3. Let $X = S^1$. Then $\pi_1(X) = H_1(X) = \mathbb{Z}$. If $f : S^1 \to S^1$ and $\deg f = k$ then f is homotopic to the polynomial map z^k. Now $L(f) = 1 - \deg f = 1 - k$ and $R(f) = \#Coker(1 - f_\#) = |1 - k|$. It turns out that

$$N(f) = |L(f)| = \begin{cases} 0, & \text{if } L(f) = 0; \\ R(f) = |1 - k|, & \text{if } L(f) \neq 0. \end{cases}$$

In 1943, W. Franz [27] (also see [40]) proved that every map $f : L(p,q) \to L(p,q)$ of the classical lens space $L(p,q)$ has the property that all fixed point classes have the same fixed point index. It follows that

$$L(f) = 0 \Rightarrow N(f) = 0$$

$$L(f) \neq 0 \Rightarrow N(f) = R(f).$$

For any $\alpha \in Cov(\eta : S^3 \to L(p,q))$, α covers $1_{L(p,q)}$ so $\deg \alpha = 1$. It follows that $\alpha \simeq 1_{S^3}$. For any lift $\alpha \tilde{f}$, α_* acts trivially on rational homology so that $L(\alpha \tilde{f})$ is constant. Note that $L(\alpha \tilde{f})$ is an integral multiple of the fixed point index of the class $\eta(Fix(\alpha \tilde{f}))$. By showing that this integral multiple is k for some constant k, we conclude that $L(f) = N(f) \cdot k$. Such a result is fairly common for a large class of spaces known as *Jiang* spaces.

In 1963, B. Jiang (see e.g. [5] or [40]) introduced a condition on the map at the fundamental group level under which all fixed point classes have the same index. For this property to hold for all maps, the Jiang condition asserts that if

$$J(X) := \mathrm{Im}(ev_* : \pi_1(X^X, 1_X) \to \pi_1(X, x_0))$$

is equal to $\pi_1(X, x_0)$ where ev is the evaluation map at x_0 then for every self-map, the fixed point classes have the same fixed point index.

Remark 1.4. The subgroup $J(X)$ is called the Jiang subgroup and is also known as the first Gottlieb group. Using Nielsen fixed point theory, D. Gottlieb [34] showed that if X is finitely dominated by a finite aspherical CW complex and $\chi(X) \neq 0$ then the center of $\pi_1(X, x_0)$ must be trivial. The subgroup $J(X)$ can also be defined as the set $\{\sigma \in Cov(\eta) | \sigma\alpha = \alpha\sigma, \forall\alpha \in Cov(\eta) \text{ and } \sigma \simeq 1_{\tilde{X}}\}$.

Definition 1.2. *A space X is called a Jiang space if $J(X) = \pi_1(X, x_0)$.*

Note that $J(X)$ is a subgroup of the center of $\pi_1(X, x_0)$. So, if X is a Jiang space then $\pi_1(X, x_0)$ is necessarily abelian.

Jiang spaces include simply connected spaces; classical and generalized lens spaces; H-spaces; coset spaces of the form G/G_0 where G is a compact Lie group and G_0 a connected closed subgroup (called homogeneous spaces).

Another technique for calculating the Nielsen number is the product formula for Nielsen numbers for fiber maps. In 1967, R. Brown [7,8] showed under certain conditions that the Nielsen number of a self-map $E \xrightarrow{f} E$ of the total space of a fibration is the product of the Nielsen number of the induced map on the base and of the Nielsen number of the restriction on a typical fiber.

$$\begin{array}{ccc} F & \xrightarrow{f_b} & F \\ \downarrow & & \downarrow \\ E & \xrightarrow{f} & E \\ {\scriptstyle p}\downarrow & & \downarrow{\scriptstyle p} \\ B & \xrightarrow{\bar{f}} & B \end{array}$$

If p is an orientable fibration, the product formula for Lefschetz numbers holds, that is, $L(f) = L(\bar{f}) \cdot L(f_b)$. In general, one can only expect some kind of addition formula $N(f) = \sum_{b \in \mathcal{E}(\bar{f})} N(f_b)$ where b varies over a set of representatives of essential fixed point classes of $\bar{f}$ (see e.g. [37]).

In 1984, D. Anosov [1] showed that if $f : N \to N$ is a self-map of a compact nil-manifold then $N(f) = |L(f)|$. This result has since generated much further research since nil-manifolds (except for tori) are not Jiang spaces but they are of *Jiang-type*, that is, either $L(f) = 0 \Rightarrow N(f) = 0$ or $L(f) \neq 0 \Rightarrow N(f) = R(f)$. In fact, Anosov's theorem asserts that the fixed point classes are either all inessential of index zero or all essential of index $+1$ or of index -1. Subsequently, it has been proved that

$$N(f) > 0 \Leftrightarrow R(f) < \infty \Leftrightarrow Fix(f_\#) = \{1\}.$$

If one of them holds then $N(f) = R(f)$ $(= |L(f)|)$(see e.g. [25]).

Recall that a nil-manifold is the coset space of a nilpotent Lie group G by a co-compact closed subgroup Γ. A classical result of A. Mal'cev says that G may be taken to be simply connected and Γ a lattice. The simplest nil-manifold that is not a torus is the following three dimensional compact manifold.

Example 1.4. Let $N = G/\Gamma$ be the three dimensional Heisenberg manifold, the coset space of the simply connected nilpotent Lie group

$$G = \left\{ \begin{pmatrix} 1 & a & b \\ 0 & 1 & c \\ 0 & 0 & 1 \end{pmatrix} |a, b, c \in \mathbb{R} \right\}$$

by the lattice

$$\Gamma = \left\{ \begin{pmatrix} 1 & a & b \\ 0 & 1 & c \\ 0 & 0 & 1 \end{pmatrix} |a, b, c \in \mathbb{Z} \right\}.$$

One can define solvmanifolds in a similar fashion: G/Γ where G is a solvable Lie group and Γ a co-compact closed subgroup. However, not all solvmanifolds G/Γ have discrete Γ. Moreover, many nice properties and fixed point results for nil-manifolds do not hold for solvmanifolds or for infra-abelian/nil/solv manifolds.

The infra-homogeneous spaces are classical aspherical manifolds. The infra-abelian manifolds and the infra-nil manifolds correspond to the flat and the almost flat Riemannian manifolds, respectively, which are precisely the finite $K(\pi, 1)$ manifolds where π is respectively in the class of finitely generated torsion-free virtually abelian (Bieberbach) groups, and virtually nilpotent groups. While Anosov's theorem does not hold in general for these infra-homogeneous spaces, similar results can be obtained under certain conditions. For instances, let $f : M \to M$ and (i) M is a solvmanifold with $R(f) < \infty$, then $N(f) = R(f)$; (ii) M is an infra-nil manifold with holonomy group Φ such that Φ does not contain a subgroup of index 2 then $N(f) = |L(f)|$ [12].

To obtain a Jiang-type result, that is, $L(f) = 0 \Rightarrow N(f) = 0$ and $L(f) \neq 0 \Rightarrow N(f) = R(f)$, one only needs to ensure that the fixed point classes have fixed point index of the same *sign*. Along this line, a family of spaces called $\mathcal{C}_0$-nilpotent spaces, where $\mathcal{C}_0$ denotes the class of finite groups, have been shown to be of Jiang-type if the center of the fundamental group of such a space has finite index. An example of such spaces is the coset space G/K of a compact connected Lie group by a finite subgroup K (see e.g. [66]).

Problem 1. Enlarge the family of spaces for which Jiang-type results hold for all self-maps.

Next, we will look at the Reidemeister trace, which contains information about both the homological indices (Lefschetz number) and the geometric data (Nielsen number).

2. Reidemeister trace and reidemeister classes – Part I

In 1936, K. Reidemeister studied Nielsen fixed point theory from the algebraic viewpoint using the universal cover. We shall define a trace-like quantity which captures both $N(f)$ and $L(f)$.

Let X be a compact connected polyhedron and $f : X \to X$ a self-map. Suppose $\tilde{f} : \tilde{X} \to \tilde{X}$ is a lift of f. First, $\tilde{X}$ inherits a simplicial structure from that of X.

The simplicial chain complex $\{C_p(\tilde{X}), \partial\}$ is defined and $C_p(\tilde{X})$ is generated by the oriented p-simplices on $\tilde{X}$.

Again, we identify $\pi = \pi_1 X \equiv Cov\eta$. If $\alpha \in \pi$ and $b \in X$ then $\alpha : \eta^{-1}(b) \to \eta^{-1}(b)$. In fact, for any $\tilde{b}_1, \tilde{b}_2 \in \eta^{-1}(b)$, there exists a unique $\alpha \in \pi$ such that $\tilde{b}_2 = \alpha(\tilde{b}_1)$. We can think of $C_p(\tilde{X})$ as a collection of π-equivariant p-simplices generated by the p-simplices of X. Since X is compact, $C_p(\tilde{X})$ is a finitely generated free $\mathbb{Z}\pi$-module. Here $\mathbb{Z}\pi$ is the integral group ring of π, i.e., a typical element is of the form $\sum n_\alpha \alpha$, $n_\alpha \in \mathbb{Z}$, $\alpha \in \pi$ and all but a finite number of the n_α's are zero. A $\mathbb{Z}\pi$ basis for $C_p(\tilde{X})$ is a collection of p-simplices $\tilde{\sigma}_1, \ldots, \tilde{\sigma}_k$, each of which corresponds to a distinct p-simplex σ_i in X. By choosing a $\mathbb{Z}\pi$ basis, the chain map

$$\tilde{f}_{\#p} : C_p(\tilde{X}) \to C_p(\tilde{X})$$

has a matrix $\mathcal{M}_p$ and the trace $\mathrm{tr}\mathcal{M}_p$ is an element of the group ring $\mathbb{Z}\pi$. By fixing the lift $\tilde{f}$, we have a homomorphism $\varphi : \pi \to \pi$. We write $\mathcal{R}(\varphi)$ as the set of orbits of the Reidemeister action

$$\sigma \bullet \alpha \mapsto \sigma \alpha \varphi(\sigma)^{-1}.$$

Let $\rho : \pi \to \mathcal{R}(\varphi)$ be the orbit map which extends linearly to $\rho : \mathbb{Z}\pi \to \mathbb{Z}\mathcal{R}(\varphi)$.

The *Reidemeister trace* [62] of f (also known as the generalized Lefschetz number) with respect to $\tilde{f}$ is given by

$$RT(f, \tilde{f}) := \sum_{q=0} (-1)^q \rho \circ \mathrm{tr}\mathcal{M}_q \qquad \in \mathbb{Z}\mathcal{R}(\varphi). \tag{2.1}$$

This "trace" is taken at the universal cover level and thus detects the fixed points of lifts of f. Suppose $\tilde{x}$ is a fixed point of $\tilde{f}$ covering the fixed point x of f. Without loss of generality, we may assume that f has finite number of fixed points each of which lies in the interior of a maximal simplex.

Since η is a local homeomorphism, we conclude that

$$index(\tilde{f}; \tilde{x}) = index(f; x).$$

Remember that lifts of the form $\gamma \tilde{f} \gamma^{-1}$ also have fixed points which project to x. Thus, by passing to $\mathcal{R}(\varphi)$, we would count the contribution to the trace exactly once.

Similar to the Lefschetz-Hopf Theorem, we have the following representation of the Reidemeister trace, due to Wecken [62].

Theorem 2.1.

$$RT(f, \tilde{f}) = \sum_{\alpha \in \mathcal{R}(\varphi)} i_\alpha \alpha \qquad i_\alpha \in \mathbb{Z}$$

where α is the Reidemeister class whose corresponding Nielsen class has fixed point index i_α.

Note that if α corresponds to an empty Nielsen class then $i_\alpha = 0$. The Wecken representation of Theorem 2.1 implies that $N(f) = \#\{\alpha | i_\alpha \neq 0\}$ and $L(f) = \sum i_\alpha$. Therefore, $RT(f, \tilde{f})$ captures information on $N(f)$ and on $L(f)$. Furthermore, the Reidemeister trace has the same type of properties as the ordinary Lefschetz number.

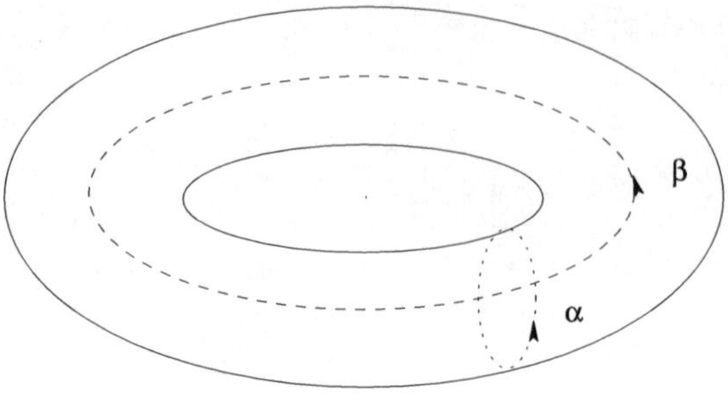

Figure 6. 2-torus

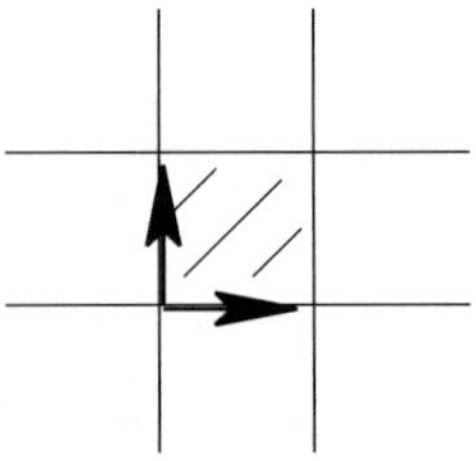

Figure 7.

It is clear that if $RT(f, \tilde{f}) \neq 0$ then $Fix(f) \neq \emptyset$. Conversely, if X is a compact manifold of dimension at least 3 then $RT(f, \tilde{f}) = 0$ implies that $N(f) = 0$ and thus $f \simeq f'$ with $Fix(f') = \emptyset$. In other words, for spaces that are *Wecken*, the Reidemeister trace $RT(f, \tilde{f})$ is a *complete* invariant.

Question. How do we compute $RT(f, \tilde{f})$?

Unlike the classical Lefschetz trace which can be computed at the homology level and is equivalent to the trace at the chain level (Hopf trace theorem), the Reidemeister trace is defined only at the chain level. In 1983, E. Fadell and S. Husseini [17] used Fox Calculus to compute $RT(f, \tilde{f})$ for surface maps.

Consider the following simple example which computes the Reidemeister trace for a map on the 2-torus.

Let $T^2 = S^1 \times S^1 = \mathbb{R}^2/\mathbb{Z}^2$. The fundamental group is given by

$$\pi_1(T^2) = \mathbb{Z} \oplus \mathbb{Z} = \langle\alpha\rangle \oplus \langle\beta\rangle.$$

Let $f : T^2 \to T^2$ be a map whose induced homomorphism on π_1 is given by φ which sends $\alpha \mapsto \beta^2\alpha^{-1}$ and $\beta \mapsto \beta\alpha^{-1}$. Consider the lift $\tilde{f} : \mathbb{R}^2 \to \mathbb{R}^2$ given by the linear map $\left(\begin{smallmatrix} -1 & -1 \\ 2 & 1 \end{smallmatrix}\right)$. Consider the cells:

$$\tilde{e}_0 = (0, 0), \qquad \tilde{e}_1^1 = (s, 0), \tilde{e}_1^2 = (0, t)$$

and the square $\tilde{e}_2 = (s, t), 0 \leq s \leq 1, 0 \leq t \leq 1$ (see Figure 7).

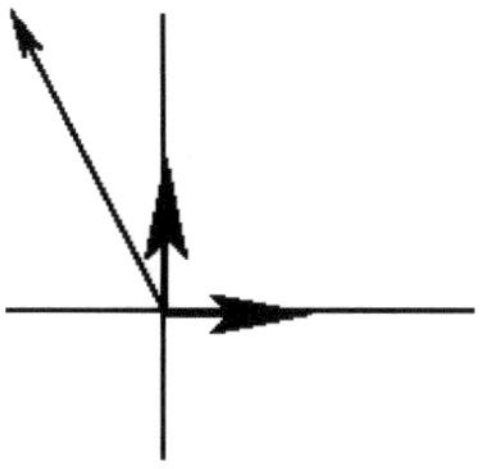

Figure 8.

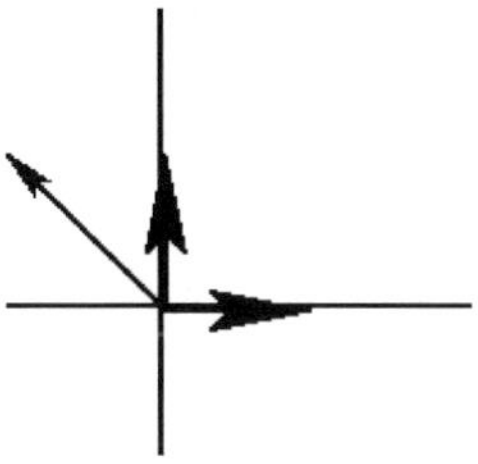

Figure 9.

Then,

$$\tilde{f}_{\#0}(\tilde{e}_0) = \tilde{e}_0;$$

$$\tilde{f}_{\#0}(\tilde{e}_1^1) = -(\beta^2\alpha^{-1})\tilde{e}_1^1 + (1+\beta)\tilde{e}_1^2;$$

$$\tilde{f}_{\#0}(\tilde{e}_1^2) = -\alpha^{-1}\tilde{e}_1^1 + \alpha^{-1}\tilde{e}_1^2;$$

$$\tilde{f}_{\#0}(\tilde{e}_2) = (\alpha^{-1} + \alpha^{-1}\beta - \alpha^{-2}\beta^2)\tilde{e}_2.$$

To see this, consider the 1-cells (see Figure 8 and Figure 9):
Since $\begin{pmatrix} -1 & -1 \\ 2 & 1 \end{pmatrix}\begin{pmatrix} 1 \\ 0 \end{pmatrix} = \begin{pmatrix} -1 \\ 2 \end{pmatrix}$, we have

$$\tilde{f}_{\#1}(\tilde{e}_1^1) = -(\beta^2\alpha^{-1})\tilde{e}_1^1 + (1+\beta)\tilde{e}_1^2.$$

Similarly, $\begin{pmatrix} -1 & -1 \\ 2 & 1 \end{pmatrix}\begin{pmatrix} 0 \\ 1 \end{pmatrix} = \begin{pmatrix} -1 \\ 1 \end{pmatrix}$, it follows that

$$\tilde{f}_{\#0}(\tilde{e}_1^2) = -\alpha^{-1}\tilde{e}_1^1 + \alpha^{-1}\tilde{e}_1^2.$$

We determine the images of the 0-cell and the 2-cell in a similar fashion.
Therefore,

$$RT(f,\tilde{f}) = [1] - (-[\beta^2\alpha^{-1}] + [\alpha^{-1}]) + ([\alpha^{-1}] + [\beta\alpha^{-1}] - [\beta^2\alpha^{-2}]). \quad (2.2)$$

Now the question is: *Is* (2.2) *the Wecken representation as in Theorem 2.1?*
Next, we need to determine the Reidemeister classes in $\mathcal{R}(\varphi)$. First, note that

$$\varphi(\alpha) = \beta^2\alpha^{-1} \quad \text{and} \quad \varphi(\beta) = \beta\alpha^{-1}.$$

Since π is abelian, if $\sigma \in \pi$ then $\sigma = \alpha^m \beta^n$ for some $m, n \in \mathbb{Z}$. Thus,

$$\begin{aligned}\sigma \alpha \varphi(\sigma)^{-1} &= \alpha^m \beta^n \alpha \varphi(\alpha^{-m} \beta^{-n}) \\ &= \alpha^m \beta^n \alpha \beta^{-2m} \alpha^m \beta^{-n} \alpha^n \\ &= \alpha^{n+1} \beta^{-2m}\end{aligned}$$

or $\sigma \varphi(\sigma)^{-1} = \alpha^n \beta^{-2m}$.

This means that the unity 1 and any element of the form $\alpha^n \beta^{\text{even}}$ are in the same Reidemeister class. Hence,

$$\mathcal{R}(\varphi) = \{[1], [\beta]\} \cong \mathbb{Z}_2.$$

Moreover,

$$[\beta^2 \alpha^{-1}] = [1] = [\alpha^{-1}] = [\alpha] = [\beta^2 \alpha^{-2}]$$

and

$$[\beta \alpha^{-1}] = [\beta].$$

We conclude that

$$RT(f, \tilde{f}) = [1] + [\beta]$$

which implies that $L(f) = 2$ and $N(f) = 2$.

In the case of an orientable surface of genus $g > 1$, $X = \sum^g$, $\pi_1(X, x_0)$ is generated by $2g$ elements satisfying one relation. Using the Fox derivative on integral group rings, E. Fadell and S. Husseini [17] computed the Reidemeister trace and showed that

$$RT(f, \tilde{f}) = \rho\left(1 - \sum_i \frac{\partial f_\#(a_i)}{\partial a_i} + A\right) \tag{2.3}$$

where $\left(\frac{\partial f_\#(a_i)}{\partial a_j}\right) = J(\varphi)$ is the Jacobian matrix and $A \in \mathbb{Z}\pi$ such that $A(\nabla R) = \varphi(\nabla R) J(\varphi)$, R being the relation.

Recall that the Fox Derivative on a free group $F_r = \langle a_1, \ldots, a_r \rangle$ of rank r is an operator $\frac{\partial}{\partial a_j} : F_r \to \mathbb{Z}F_r$ for each j, $1 \leq j \leq r$ such that

$$\frac{\partial}{\partial a_i}(a_j) = \delta_{ij};$$

$$\frac{\partial}{\partial a_i} 1 = 0;$$

$$\frac{\partial}{\partial a_i} uv = \frac{\partial}{\partial a_i} u + u \frac{\partial}{\partial a_i} v.$$

It follows that

$$\frac{\partial}{\partial a_i} w^{-1} = -w^{-1} \frac{\partial}{\partial a_i} w, \quad \text{and} \quad \frac{\partial}{\partial a_i}(a_i^n) = 1 + a_i + \cdots + a_i^{n-1}, \quad n \geq 1.$$

3. Reidemeister trace and reidemeister classes – Part II

The most difficult part of the computation is the ability to express the Reidemeister trace in the Wecken representation, i.e., the ability to distinguish Reidemeister classes. The problem of distinguishing Reidemeister classes is equivalent to the twisted conjugacy problem, namely, given a group endomorphism $\varphi : \pi \to \pi$ and two elements α, β, is there an algorithm to determine if $\beta = \sigma\alpha\varphi(\sigma)^{-1}$ for some $\sigma \in \pi$?

There are some *ad hoc* methods that can be used to determine if two elements in the fundamental group belong to the same Reidemeister class. The first is to use abelianization.

Example 3.1. We now revisit Example 1.1. The induced homomorphism on the free group on two generators $\pi = \langle a, b\rangle$ is given by $a \mapsto a^{-1}$ and $b \mapsto b^2$. As we have already seen that

$$\frac{\partial}{\partial a}\varphi(a) = -a^{-1} \quad \text{and} \quad \frac{\partial}{\partial b}\varphi(b) = 1 + b$$

and thus

$$RT(f, \tilde{f}) = \rho(a^{-1} - b) = [a^{-1}] - [b].$$

Hence, $L(f) = 0$. To show that $[a^{-1}] \neq [b]$, we consider φ_{Ab}, the induced homomorphism on the abelianization $\pi_{Ab} = \langle\bar{a}\rangle \oplus \langle\bar{b}\rangle$ where $\bar{a}, \bar{b}$ are the respective images of a, b in π_{Ab}. If $[a^{-1}] = [b]$ then $\bar{a}^{-1}$ and $\bar{b}$ would be φ_{Ab}-conjugated. In other words, there exists $\bar{\sigma} \in \pi_{Ab}$ such that $\bar{b} = \bar{\sigma} + \bar{a}^{-1} - \varphi_{Ab}(\bar{\sigma})$. If

$$\bar{\sigma} = \begin{pmatrix} m \\ n \end{pmatrix} \in \mathbb{Z}^2, \text{ then } \varphi_{Ab}(\bar{\sigma}) = \begin{pmatrix} -m \\ 2n \end{pmatrix} \text{ since } \varphi_{Ab} = \begin{pmatrix} -1 & 0 \\ 0 & 2 \end{pmatrix}.$$

However,

$$\bar{b} = \begin{pmatrix} 0 \\ 1 \end{pmatrix} = \begin{pmatrix} m \\ n \end{pmatrix} - \begin{pmatrix} 1 \\ 0 \end{pmatrix} - \begin{pmatrix} -m \\ 2n \end{pmatrix} = \begin{pmatrix} 2m - 1 \\ -n \end{pmatrix}.$$

This implies that $\bar{\sigma} = \begin{pmatrix} 1/2 \\ -1 \end{pmatrix}$ which is a contradiction. It follows that $N(f) = 2$.

Example 3.2 (due to P. Staecker [54]). Let $\varphi : \pi = F_3 \to F_3$ be given by $a \mapsto a^{-2}c$, $b \mapsto b^{-1}$, and $c \mapsto bca^{-1}$ where $F_3 = \langle a, b, c\rangle$ is the free group on three generators. Using Fox Calculus, we obtain

$$RT(f, \tilde{f}) = \rho(1 + a^{-1} + a^{-2} + b^{-1} - b).$$

Thus, there are at most 5 distinct essential Nielsen classes, i.e., $1 \leq N(f) \leq 5$. Since $\varphi(b) = b^{-1}$, it follows that $[b] = [b^{-1}]$ and thus

$$RT(f, \tilde{f}) = \rho(1 + a^{-1} + a^{-2}).$$

Now, we know that $1 \leq N(f) \leq 3$. In the abelianization π_{Ab}, the induced map φ_{Ab} is given by the matrix

$$\begin{pmatrix} -2 & 0 & 1 \\ 0 & -1 & 0 \\ -1 & 1 & 1 \end{pmatrix}.$$

It is easy to check that $[1] \neq [a^{-1}]$ because $\bar{1}$ and $\overline{a^{-1}}$ are not φ_{Ab}-twisted conjugated in π_{Ab}. Similarly, one shows that $[a^{-1}] \neq [a^{-2}]$. Now, we know that $2 \leq N(f) \leq 3$.

At this point, this is all we know from abelianization. If one considers the nilpotent quotient $\pi/\gamma_2(\pi)$, then using Hall's normal form (see e.g. p.343 of [49]), one can deduce that $[1] \neq [a^{-2}]$ and hence

$$RT(f, \tilde{f}) = \rho(1 + a^{-1} + a^{-2}) = [1] + [a^{-1}] + [a^{-2}].$$

is indeed the Wecken representation so that $N(f) = 3$.

Problem 2. Find conditions under which there is an algorithm to distinguish Reidemeister classes using nilpotent quotients.

Next, we briefly describe an algorithm for computing the Nielsen number for certain maps on surfaces with boundary, due to J. Wagner (1999) [61].

Let $f : X \to X$ be a self-map on a surface X with boundary so that $\pi = \pi_1(X, x_0)$ is a free group on r generators. The Reidemeister trace becomes

$$RT(f, \tilde{f}) = \rho\left(1 - \sum_i \frac{\partial f_\#(a_i)}{\partial a_i}\right)$$

since $A = 0$ in (2.3). We let $T = \left(1 - \sum_i \frac{\partial f_\#(a_i)}{\partial a_i}\right)$. Wagner's algorithm gives a method, under certain conditions, to determine if elements in T belong to the same Reidemeister class.

Remark 3.1. We should point out that the Reidemeister action used in [17,61,45], and others is a *right* action $\alpha \cdot \sigma \mapsto \varphi(\sigma)^{-1}\alpha\sigma$. By converting this right action to the left action defined above, we have a bijection between the Reidemeister traces defined on the two different actions, that is, $[\alpha] \leftrightarrow [\alpha^{-1}]$.

Example 3.3. Let $\pi = F_4$ and $f_\# = \varphi$ be given by

$$a \mapsto a^2b^{-1}a,\ b \mapsto b^{-2},\ c \mapsto bc^2ba^{-2},\ d \mapsto d^2b^{-1}a.$$

Then,

$$T = 1 - 1 - a - a^2b^{-1} + b^{-1} + b^{-2} - b - bc - 1 - d.$$

Now order the terms by the order in which the elements appear in T. The first one is 1 and is called W_0. If W_i is the element from $\frac{\partial f_\#(a_j)}{\partial a_j}$, we write $\overline{W_i} = f_\#(a_j)^{-1}W_i a_j$ and $\overline{W_0} = 1$. Here we use the right action $\alpha \cdot \sigma \mapsto \varphi(\sigma)^{-1}\alpha\sigma$ instead.

Clearly, $[W_i] = [W_j]$ in $\mathbb{Z}\mathcal{R}(\varphi)$ if $W_i = W_j$, $W_i = \overline{W_j}$, or $\overline{W_i} = \overline{W_j}$. In such a situation, we say that $W_i \overset{\text{dir}}{\simeq} W_j$ (directly related). Here, we have the following table:

order	generator	coefficient	W_i	$\overline{W_i}$
0		$+1$	1	1
1	a	-1	1	$a^{-1}ba^{-1}$
2	a	-1	a	$a^{-1}b$
3	a	-1	a^2b^{-1}	1
4	b	$+1$	b^{-1}	b^2
5	b	$+1$	b^{-2}	b
6	c	-1	b	$a^2b^{-1}c^{-1}$
7	c	-1	bc	a^2b^{-1}
8	d	-1	1	$a^{-1}bd^{-1}$
9	d	-1	d	$a^{-1}b$

In each of the following sets, the elements are Reidemeister equivalent: $\{W_0, W_1, W_3, W_7, W_8\}$, $\{W_2, W_9\}$, $\{W_5, W_6\}$. It is easy to see that $[W_0]$ has index -3, $[W_2]$ has index -2, $[W_5]$ has index 0, and $[W_4]$ has index 1. A result of J. Wagner [ibid] implies that $RT(f, \tilde{f})$ has four distinct classes with the respective fixed point index so that $L(f) = -4$ and $N(f) = 3$.

An endomorphism $\varphi : F_r \to F_r$ is *W-characteristic* if whenever W_x and W_y are Reidemeister equivalent there exists a sequence $W_{k_1}, \ldots, W_{k_s}$ such that $W_x \overset{\text{dir}}{\simeq} W_{k_1}$, $W_{k_s} \overset{\text{dir}}{\simeq} W_y$, and $W_{k_t} \overset{\text{dir}}{\simeq} W_{k_{t+1}}$, for $t = 1, \ldots, s-1$.

J. Wagner [ibid] gave conditions on φ under which φ becomes W-characteristic and Reidemeister equivalence can be determined. Subsequently, P. Yi [69] and more recently S. Kim [45] improved on Wagner's algorithm so that more classes of endomorphisms on F_r can be included in the family of maps for which Reidemeister equivalence can be determined.

Remark 3.2. One can adapt Wagner's algorithm with respect to the *left* action. We modify the above table of W_i and $\overline{W_i}$ by simply inverting the columns of the generators and of the W_i's and defining $\overline{W_i} = a_j^{-1} W_i f_\#(a_j)$.

In 2006, O. Bogopolski, A. Martino, O. Maslakova, and E. Ventura [3] proved that the twisted conjugacy problem for automorphisms of free groups of finite rank is solvable, i.e., there is an algorithm that decides whether α and β are φ-twisted conjugated in π. The underlying idea is the use of the algebraic mapping torus.

Let π be a finitely generated group, $\varphi \in \mathrm{Aut}(\pi)$. The mapping torus is the semidirect product

$$G = \pi \rtimes_\varphi \mathbb{Z} = \langle \alpha, t | \alpha \in \pi, t \in \mathbb{Z}, t\alpha t^{-1} = \varphi(\alpha) \rangle.$$

This is an HNN-extension, a well-known construction in group theory.

Suppose $\alpha, \beta \in \pi$ and $\beta = \sigma\alpha\varphi(\sigma)^{-1}$ for some $\sigma \in \pi$. Then,

$$\beta t = \sigma\alpha\varphi(\sigma)^{-1}t = \sigma(\alpha t)\sigma^{-1}.$$

Conversely, if $\beta t = \gamma\alpha t\gamma^{-1}$, then it follows easily that α and β are φ-conjugated in π.

Remark 3.3. If φ is not an automorphism, we can still use the mapping torus to distinguish Reidemeister classes, i.e., if αt and βt are *not* conjugate to each other in G then α and β are not φ-conjugated in π.

Bogopolski et al. [3] showed that the conjugacy problem is solvable for the mapping tori $F_r \rtimes_\varphi \mathbb{Z}$, and thus the φ-twisted conjugacy problem is solvable for automorphism φ on F_r. Their result relies on the following theorems.

Theorem 3.1 ([50]). *There exists an algorithm to compute a finite generating set for the fixed subgroup of an automorphism of a free group of finite rank.*

Theorem 3.2 ([4]). *Given $u, v \in F_r$ and $\varphi \in \mathrm{Aut}(F_r)$, it is decidable whether there exists an integer k such that $\varphi^k(u)$ is conjugate to v.*

Problem 3. While [3] shows that the twisted conjugacy problem is solvable for automorphisms of free groups of finite rank, its algorithm is not easy to implement whereas Wagner's method is easy to use. Find an effective algorithm for solving the twisted conjugacy problem for automorphisms of finitely generated free groups.

Remark 3.4. The Reidemester trace also arose in the context of Bass' conjectures. Recent work of A. Berrick, I. Chatterji, and G. Mislin [2] gave a geometric interpretation based upon an observation of R. Geoghegan relating the Bass Conjectures and homotopy idempotent maps on manifolds. The (strong) Bass conjecture states that for any discrete group π, the Hattori-Stallings trace of a finitely generated projective module over $\mathbb{Z}\pi$ is supported on the identity component only. A weaker form of the conjecture asserts that the sum of the coefficients of the non-identity components be zero. In [2], the strong Bass conjecture is shown to be equivalent to the statement that for every homotopy idempotent self-map f of a closed smooth orientable n-manifold, $n \geq 3$, $N(f) = 1$. Furthermore, it was also shown in [2] that the weak Bass conjecture is equivalent to the statement that every pointed homotopy idempotent self-map inducing the identity on the fundamental group has its Lefschetz number coincide with the L^2-Lefschetz number of the induced map on the universal cover.

4. R_∞ property – Part I

We begin with the following question: Given $\varphi \in \mathrm{Aut}(\pi)$, when is $R(\varphi)$ finite? We will give motivation from three different viewpoints.

An old question in group theory asks whether every infinite group π must have an infinite number of conjugacy classes, that is, if $R(id_\pi) = \infty$. The answer is yes if π is an infinite abelian group. In 1949, Higman-Neumann-Neumann [38] constructed an infinitely generated group with only a finite number of conjugacy classes. Later, S. Ivanov [39] constructed one such example where π is finitely generated. More recently, D. Osin [52] exhibited a finitely generated group π in which any two non-trivial elements are conjugate in π. The finiteness of $R(\varphi)$ is a generalization of the same problem with $\varphi = id_\pi$.

From the fixed point theory point of view, Jiang-type results are very useful in terms of computing the Nielsen number. For such spaces, either $N(f) = 0$ or $N(f) = R(f)$. If $R(f) = \infty$, then $N(f) = 0$ and in many cases, f is deformable to be fixed point free. If in addition X is a compact manifold of dimension at least 5, then a result of M. Kelly (1995) [43] implies that every homeomorphism f is isotopic to a fixed point free homeomorphism. But before we can establish such Jiang-type results, we must need to know whether $R(f)$ is finite or not for the equality $R(f) = N(f)$ to hold. This kind of question already arose concerning self-maps of nil-manifolds. The finiteness of $R(f)$ has further implication for other spaces like solvmanifolds and for Nielsen coincidence theory.

From the representation theory point of view, the classical Burnside-Frobenius theorem says that the number of classes of irreducible representations is equal to the number of conjugacy classes of elements of π. Let $\widehat{\pi}$ be the *unitary dual* of π, i.e. the set of equivalence classes of unitary irreducible representations of π. If $\varphi \in \mathrm{Aut}(\pi)$, it induces a map $\widehat{\varphi} : \widehat{\pi} \to \widehat{\pi}$ given by $\widehat{\varphi}(\rho) = \rho \circ \varphi$. Thus, by the Burnside-Frobenius theorem, if φ is the identity automorphism of any finite group π, then we have $R(\varphi) = \#Fix(\widehat{\varphi})$ where $Fix(\widehat{\varphi})$ is the set of fixed points of $\widehat{\varphi}$. For a finite group π, A. Fel'shtyn and R. Hill [24] showed that $R(\varphi)$ is equal to the dimension of the space of twisted invariant functions on this group. Hence, $R(\varphi)$ coincides with the number of fixed points of $\widehat{\varphi}$. Recently A. Fel'shtyn and E. Troitsky [26] show that the twisted Burnside-Frobenius theorem holds for virtually polycyclic groups if the Reidemeister number is finite. Denote by $\widehat{\pi}_f$ the subset of the unitary dual $\widehat{\pi}$ related to finite-dimensional representations and by $F_f(\varphi)$ the number of fixed points of $\widehat{\varphi}_f$ on $\widehat{\pi}_f$. Then $R(\varphi) = F_f(\varphi)$ if one of these numbers is finite.

In 1994, A. Fel'shtyn and R. Hill [24] conjectured that for a finitely generated torsion-free group π with exponential growth (see p.180 of [36] for definition), any injective endomorphism $\varphi : \pi \to \pi$ will have $R(\varphi) = \infty$.

Definition 4.1. *A finitely generated group G is said to have the property R_∞ if for any $\varphi \in Aut(G)$, $R(\varphi) = \infty$.*

In 2000, G. Levitt and M. Lustig [48] used geometric methods to show that finitely generated torsion-free non-elementary Gromov hyperbolic groups have the R_∞ property. A. Fel'shtyn [20] extended Levitt-Lustig's result by removing the torsion-free hypothesis. Since Z. Sela [53] showed that finitely generated torsion-free freely indecomposable Gromov hyperbolic groups are co-Hopfian, it follows that the Fel'shtyn-Hill conjecture holds for this class of groups.

In 2003, D. Gonçalves and the author [29] gave examples of non-Gromov hyperbolic groups of exponential growth in which automorphisms with finite Reidemeister number exist. Since then, more examples of groups have been found to have the R_∞ property.

Groups with property R_∞ include

(1) Baumslag-Solitar groups $BS(m,n) = \langle a, b | ba^m b^{-1} = a^n \rangle$ except for $BS(1,1)$. [21]
(2) Generalized Baumslag-Solitar groups, that is, finitely generated groups which act on a tree with all edge and vertex stabilizers infinite cyclic. [47]
(3) Lamplighter groups $\mathbb{Z}_n \wr \mathbb{Z}$ iff $2|n$ or $3|n$. [32]
(4) Some groups of the form $\mathbb{Z}^2 \rtimes \mathbb{Z}$. [29]
(5) Non-elementary Gromov hyperbolic groups. [48,20]
(6) The solvable generalization Γ of $BS(1,n)$ given by the short exact sequence $1 \to \mathbb{Z}[\frac{1}{n}] \to \Gamma \to \mathbb{Z}^k \to 1$ as well as any group quasi-isometric (c.f. Definition 5.1) to Γ. [56]
(7) Mapping class groups of surfaces. [22]
(8) Relatively hyperbolic groups. [22]
(9) Any group with a *characteristic* and non-elementary action on a Gromov hyperbolic space. [55,22]

These results fall into two categories: those which show the property holds using a group presentation, and those which show that the R_∞ property is geometric in some way, whether invariant under quasi-isometry, or dependent on an action of a group on a particular space. The final statement on the list is the most general, relying only on the existence of a certain group action. This was proven using similar methods in [55] and independently in the Appendix of [22], and provides a wealth of examples.

The Fel'shtyn-Hill conjecture seems to suggest that the growth rate is a contributing factor in the infiniteness of $R(\varphi)$. Indeed, if $\pi = \mathbb{Z}^k$, then π does not have R_∞ since one can easily find an automorphism $\varphi : \pi \to \pi$ such that $\det(I - \varphi) \neq 0$ which implies that $R(\varphi) < \infty$. It is then natural to ask if one can find a finitely generated torsion-free nilpotent group with the R_∞ property. If there is, then we know that R_∞ does *not* depend on the growth type for such a group has polynomial growth.

In 1970, J. Dyer [13] constructed some nilpotent Lie algebras which have the property that every automorphism is unipotent. This implies that the induced homomorphism on the abelianization is the identity. This should imply that every automorphism of the corresponding nilpotent group has the property that the induced automorphism on the abelianization is the identity. An immediate consequence of this is the fact that every automorphism ϕ of such a nilpotent group has an infinite number of ϕ-twisted conjugacy classes and thus the group has the R_∞ property. This has motivated the following problem: classify all finitely generated torsion-free nilpotent groups with the R_∞ property.

Given r a positive integer and k either an integer or ∞, consider the free nilpotent group $G(r,k) = F_r/\Gamma_{k+1}(F_r)$ of rank r and nilpotency class k, where for $k = \infty$ we have $G(r,\infty) = F_r$ because F_r is residually nilpotent. For which values of r and k(including $k = \infty$) does the group G have the R_∞ property?

Let us denote $\Gamma_n(F_2)$ simply by Γ_n. Using commutator analysis on free groups, Witt's Formula calculates the ranks of the abelian groups Γ_i/Γ_{i+1}. In particular, we have: $rk(\Gamma_2/\Gamma_3) = 1$, $rk(\Gamma_3/\Gamma_4) = 2$, $rk(\Gamma_4/\Gamma_5) = 3$, $rk(\Gamma_5/\Gamma_6) = 6$, $rk(\Gamma_6/\Gamma_7) = 9$, $rk(\Gamma_7/\Gamma_8) = 18$, $rk(\Gamma_8/\Gamma_9) = 30$.

D. Gonçalves and the author [33] recently showed the following.

Theorem 4.1. *If $\phi : G(2,8) \to G(2,8)$ is an automorphism then $R(\phi) = \infty$.*

The idea of the proof is to construct a non-trivial element in $\Gamma_8/\Gamma_9 \subset G(2,8)$ which is fixed by the restriction of any automorphism $\phi : G(2,8) \to G(2,8)$, using the following commutative diagram.

$$\begin{array}{ccccccccc} 0 & \longrightarrow & \Gamma_8/\Gamma_9 & \longrightarrow & F_2/\Gamma_9 & \longrightarrow & F_2/\Gamma_8 & \longrightarrow & 1 \\ & & \phi' \downarrow & & \hat{\phi} \downarrow & & \bar{\phi} \downarrow & & \\ 0 & \longrightarrow & \Gamma_8/\Gamma_9 & \longrightarrow & F_2/\Gamma_9 & \longrightarrow & F_2/\Gamma_8 & \longrightarrow & 1 \end{array}$$

Since Γ_8/Γ_9 is finitely generated free abelian, the existence of a non-trivial element fixed by ϕ' implies that $R(\phi') = \infty$. Then one concludes that $R(\phi) = \infty$.

More generally, we have the following result.

Theorem 4.2 ([33]). *Let $G(2,k)$ be the free nilpotent group on 2 generators of nilpotency class k. If $k \geq 9$ (including the case $k = \infty$) then for any automorphism $\phi : G(2,k) \to G(2,k)$, we have $R(\phi) = \infty$, i.e., $G(2,k)$ has the R_∞ property.*

Remark 4.3. When $k = \infty$, $G(2,\infty) = F_2$. Thus, the proof of Theorem 4.2 provides a purely algebraic proof of the fact that F_2 has the R_∞ property, a consequence of a result proven by Levitt and Lustig [48] using techniques from geometric group theory.

Problem 4. Find an algebraic proof for the R_∞ property for free groups of finite rank. Determine whether free groups of infinite rank must also have the R_∞ property.

Note that the Hirsch length of F_2/Γ_9 is equal to

$$rk(F_{2Ab}) + \sum_{i=2}^{8} rk(\Gamma_i/\Gamma_{i+1}) = 71.$$

It follows that there exists a 71-dimensional compact nil-manifold such that every homeomorphism f has $R(f) = \infty$. Since nil-manifolds are Jiang-type, it follows that every such homeomorphism will have $N(f) = 0$ and hence is isotopic to a fixed point free homeomorphism.

Question. What is the smallest k for which there is a compact nil-manifold of dimension k whose fundamental group has the R_∞ property?

In dimension 2, there is one such space, namely the 2-torus T^2 whose fundamental group $\mathbb{Z}^2$ does not have the R_∞ property. In dimension 3, one can easily classify them and none of them can have R_∞.

Example 4.1. [33] Let $F_2 = \langle x, y \rangle$ and $B = [x, y]$, the commutator of x and y. Consider the group

$$G = \langle x, y | \Gamma_4, [B, y] \rangle.$$

Then, G has the R_∞ property.

Note that G is *not* free nilpotent and the Hirsch length of G is 4. Using this G, one can construct a finitely generated torsion-free nilpotent group $G \times \mathbb{Z}^n$ of Hirsch length $n + 4$ with the R_∞ property. Topologically, we obtain the following.

Theorem 4.3 ([33]). *For any $n \geq 4$, there exists a compact nil-manifold M with* $\dim M = n$ *such that every homeomorphism $f : M \to M$ is homotopic to a fixed point free map. Moreover, if $n \geq 5$ then f is isotopic to a fixed point free homeomorphism.*

Question. What is the relationship between R_∞ and the nilpotency class?

Clearly, if the nilpotency class is 1, then the group is abelian and thus does not have R_∞. We now construct a group of nilpotency class 2 which has the R_∞ property.

Example 4.2 ([33]). Consider the free nilpotent group $G(4, 2)$ on 4 generators x, y, z, w of nilpotency class 2. Note that $\Gamma_2(G(4, 2))/\Gamma_3(G(4, 2))$ is a free abelian group of rank 6 in which the cosets defined by the commutators $[x, y]$, $[x, z]$, $[x, w]$, $[y, z]$, $[y, w]$, $[z, w]$ form a basis. Take the quotient by the subgroup generated by $[x, z]$, $[x, w]$, $[y, z]$. This group has the R_∞ property.

In contrast to the last example where the group is not free nilpotent, we will next show that the *free* nilpotent groups of nilpotency class 2 do not have the R_∞ property.

Example 4.3 ([33]). We now show that the free nilpotent groups $G(r, 2)$ does not have the R_∞ property. It suffices for each r to exhibit an automorphism $\phi : G(r, 2) \to G(r, 2)$ such that $R(\phi) < \infty$.

For $r = 2$, consider the automorphism given by $\phi(x) = x^2 y$ and $\phi(y) = x^5 y^2$. The matrix of the induced automorphism on the abelianization $\mathbb{Z} \oplus \mathbb{Z}$ has two eigenvalues $\lambda_1 = 2 + \sqrt{5}$, $\lambda_2 = 2 - \sqrt{5}$. It follows that $Fix(\phi_{Ab}) = \{1\}$ so that $R(\phi_{Ab}) < \infty$. The induced automorphism ϕ' on $\Gamma_2/\Gamma_3 = \mathbb{Z}$ is multiplication by -1 so that $R(\phi') < \infty$. Since we have a central extension we have $R(\phi) = R(\phi_{Ab})R(\phi')$ and the result follows.

For $r = 3$, let $\phi(x) = x^2 y$, $\phi(y) = x^5 y^2$ and $\phi(z) = z^{-1}$. The matrix of the induced automorphism $\phi' : \Gamma_2/\Gamma_3 \to \Gamma_2/\Gamma_3$ where $\Gamma_2/\Gamma_3 = \mathbb{Z}^3$, has eigenvalues, $\{\lambda_1\lambda_2, -\lambda_1, -\lambda_2\}$. Since ϕ' does not have 1 as an eigenvalue and thus $Fix(\phi') = \{1\}$ so that $R(\phi') < \infty$.

Let $r > 3$. If r is even, define $\phi : (\mathbb{Z} \oplus \mathbb{Z})^{r/2} \to (\mathbb{Z} \oplus \mathbb{Z})^{r/2}$ as a direct sum of the automorphism $\phi : \mathbb{Z} \oplus \mathbb{Z} \to \mathbb{Z} \oplus \mathbb{Z}$ defined above. If r is odd, define

$$\phi : (\mathbb{Z} \oplus \mathbb{Z})^{r-1/2} \oplus \mathbb{Z} \to (\mathbb{Z} \oplus \mathbb{Z})^{r-1/2} \oplus \mathbb{Z}$$

as a direct sum of the automorphism $\phi : \mathbb{Z} \oplus \mathbb{Z} \to \mathbb{Z} \oplus \mathbb{Z}$ defined above and the automorphism given by multiplication by -1 in the last coordinate. It is easy to see that the product of any two eigenvalues of ϕ for either r even or r odd cannot be 1. Therefore $\phi' : \Gamma_2/\Gamma_3 \to \Gamma_2/\Gamma_3$ does not have 1 as an eigenvalue so that $Fix(\phi') = \{1\}$ and the result follows.

Question. What is the smallest k for which there is a k-dimensional compact manifold whose fundamental group has the R_∞ property?

It turns out to be 2. In fact, the Klein Bottle K is one such manifold. Furthermore, we have

Theorem 4.4 ([33]). *The group $\pi_1(K)\times\mathbb{Z}^n$ has the R_∞ property for all integer $n \geq 0$.*

The fundamental group of K is $\mathbb{Z} \rtimes \mathbb{Z}$. However, this is *not* the simplest group with the R_∞ property.

Theorem 4.5 ([33]). *The group $\mathbb{Z} \rtimes \mathbb{Z}_2$ has the R_∞ property.*

Note that $\mathbb{Z} \rtimes \mathbb{Z}_2$ has torsion so it cannot be the fundamental group of a finite aspherical manifold while K is an infra-nil-manifold. All the groups shown here that possess the R_∞ property are virtually nilpotent and hence of polynomial growth.

Remark 4.2. It is well-known that there is a one-to-one correspondence between compact flat manifolds and finitely generated Bieberbach groups, that is, compact Riemannian manifolds with zero sectional curvatures are precisely the finite dimensional aspherical manifolds with finitely generated torsion-free virtually abelian fundamental groups. It is natural to ask whether one can obtain a result similar to Theorem 4.3 for flat manifolds. One of the main differences between nil-manifolds and infra-nil-manifolds (or flat manifolds) is that $N(f) = 0$ iff $R(f) = \infty$ for self-maps on nil-manifolds whereas there are maps on flat manifolds where $R(f) = \infty$ but $N(f) > 0$. Even for the Klein bottle K, one can find a homeomorphism with $N(f) > 0$ and $R(f) = \infty$. Thus, the groups $\pi_1(K) \times \mathbb{Z}^n$ do not provide examples of flat manifolds on which all homeomorphisms have zero Nielsen numbers. On the other hand, if G is a torsion-free finite extension of a nilpotent group N with R_∞ property such that if N is characteristic then G will have the R_∞ property.

Problem 5. Determine the R_∞ property for crystallographic groups and for virtually nilpotent groups.

Next we turn to another generalization of nilpotent groups. Nilpotent spaces have proven to be useful in homotopy theory and in algebraic topology in general. A family $\mathcal{C}$ of groups is a *class of groups* if given a short exact sequence of groups

$$1 \to A \to B \to C \to 1$$

then $A, C \in \mathcal{C}$ if and only if $B \in \mathcal{C}$. Given a class $\mathcal{C}$, a group G is said to be $\mathcal{C}$-nilpotent if $\Gamma_n(G) \in \mathcal{C}$ for some positive integer n where $\Gamma_i(G)$ is the i-th commutator in the lower central series of G. More generally, a group π acts on a group G $\mathcal{C}$-nilpotently if $\Gamma_n^\pi(G) \in \mathcal{C}$ for some positive integer n where $\Gamma_n^\pi(G)$ is the smallest π-subgroup that contains $[G, \Gamma_{n-1}^\pi(G)]$ and the set $\{(\alpha \cdot g)g^{-1} | \alpha \in \pi, g \in \Gamma_{n-1}^\pi(G)\}$. Thus, a space X is said to be $\mathcal{C}$-nilpotent if $\pi_1(X, x_0)$ is $\mathcal{C}$-nilpotent and the action of $\pi_1(X, x_0)$ on $\pi_k(X)$ is $\mathcal{C}$-nilpotent (see also [59]). Recall that $\mathcal{C}_0$ denotes the class of finite groups.

Let G be a finitely generated group. For any $\varphi \in \mathrm{Aut}(G)$ and any positive integer n, we have a commutative diagram.

$$\begin{array}{ccccccccc} 0 & \longrightarrow & \Gamma_n(G) & \longrightarrow & G & \longrightarrow & G/\Gamma_n(G) & \longrightarrow & 1 \\ & & \downarrow{\scriptstyle \varphi'} & & \downarrow{\scriptstyle \varphi} & & \downarrow{\scriptstyle \bar{\varphi}} & & \\ 0 & \longrightarrow & \Gamma_n(G) & \longrightarrow & G & \longrightarrow & G/\Gamma_n(G) & \longrightarrow & 1 \end{array} \tag{4.1}$$

It follows that if $G/\Gamma_n(G)$ has the R_∞ property then so does G. Since $N_n = G/\Gamma_n(G)$ is nilpotent, the torsion elements form a subgroup T_n, and T_n is a characteristic subgroup of $G/\Gamma_n(G)$. It follows that if N_n/T_n has the R_∞ property then N_n also has the R_∞ property.

Thus, we have the following useful result that can be used to construct more examples of groups with R_∞ property.

Proposition 4.1 ([33]). *Let G be a finitely generated group and T_k be the torsion subgroup of the nilpotent group $N_k = G/\Gamma_k(G)$. If for some positive integer n the group N_n/T_n has the R_∞ property then G also has the R_∞ property.*

We now combine our earlier results to enlarge the family of groups with the R_∞ property and also the family of compact manifolds which have fundamental groups with the R_∞ property.

Example 4.4 ([33]). Let Σ^3 be the Poincaré 3-sphere with $\pi_1(\Sigma^3) = Icos$, the binary icosahedral group of order 120. Suppose M is the 4-dimensional $K(\pi, 1)$ nil-manifold where π is the torsion-free nilpotent group of nilpotency class 3 as in Example 4.1. Take $X = \Sigma^3 \times M$. Now, $G = \pi_1(X, x_0)$ is $\mathcal{C}_0$-nilpotent, $\Gamma_4(G) = Icos$, and $G/\Gamma_4(G) = \pi$. Since π has the R_∞ property (and the torsion subgroup of π is trivial), it follows from Proposition 4.1 that G has the R_∞ property. More generally, one can obtain similar examples by replacing $Icos$ with any finite perfect group.

We can show more for homeomorphisms of the space X in the last example, although it is not clear if X is a Jiang-type space. In fact, we obtain the following result, similar to Theorem 4.3.

Theorem 4.6 ([33]). *For any $n \geq 7$, there exists a compact $\mathcal{C}_0$-nilpotent manifold M with $\dim M = n$ such that every homeomorphism $f : M \to M$ is isotopic to a fixed point free homeomorphism.*

Proof. Let $n \geq 7$ and $M = \Sigma^3 \times N^{n-3}$ where N^k is the compact nil-manifold with fundamental group of Hirsch length k as constructed in Example 4.1. This M is the desired manifold.

Remark 4.3. Although the space M in Theorem 4.6 is a $\mathcal{C}_0$-nilpotent space, its fundamental group has a center of infinite index so we cannot conclude that M is of Jiang-type. We should point out that one can construct similar manifolds M with the desired property by replacing Σ^3 with any compact manifold L with finite $\pi_1(L)$. Note that if $\dim L \leq 2$, $\pi_1(L)$ is necessarily abelian so that $\pi_1(M)$ would be nilpotent, possibly with torsion.

Remark 4.4. If H is a finite subgroup of a compact connected Lie group G, then the coset space $X = G/H$ is $\mathcal{C}_0$-nilpotent and its fundamental group has a center of finite index, hence is of Jiang-type. However, these spaces do not provide examples like Example 4.4 because $[\pi_1(X, x_0), \pi_1(X, x_0)]$ is also finite. It follows that $\pi_1(X, x_0)/[\pi_1(X, x_0), \pi_1(X, x_0)] = \pi_1(X, x_0)_{Ab}$ is finitely generated abelian, therefore does not have R_∞, and hence Proposition 4.1 is not applicable.

5. R_∞ property – Part II

In this last section, we will focus on some geometric techniques in the study of the R_∞ property of groups. We begin by considering groups as geometric objects. The basic reference for geometric group theory is [36].

Let Γ be a finitely generated group with a generating set $S = \{g_1, \dots, g_k\}$ so that Γ is a quotient of F_k, the free group on k generators, that is, there exists a surjective homomorphism $\pi : F_k \to \Gamma$. Given $\gamma \in \Gamma$, the (word) *length*

$$\ell_S(\gamma) = \min\{n | \exists \text{ a sequence } s_1, \dots, s_n \in S \cup S^{-1} \text{ such that } \gamma = \pi(s_1 s_2 \dots s_n)\}.$$

Then Γ admits a (word) metric d_S given by

$$d_S(\gamma_1, \gamma_2) := \ell_S(\gamma_1^{-1}\gamma_2).$$

Now Γ is a metric space and Γ acts on Γ (space) from the left by isometries.

The associated *Cayley graph* $Cay(\Gamma, S)$ is the graph with vertex set Γ in which γ_1, γ_2 are the ends of an edge iff $d_S(\gamma_1, \gamma_2) = 1$. Identifying each edge with $[0, 1]$ we can make $Cay(\Gamma, S)$ into a metric space and Γ acts on $Cay(\Gamma, S)$ (on the left) isometrically. The inclusion $\Gamma \hookrightarrow Cay(\Gamma, S)$ is an isometric embedding.

Example 5.1. Let $\Gamma = \mathbb{Z}$ and $S = \{1\}$. Then $Cay(\Gamma, S)$ is shown in Figure 10. If we use another generating set $S' = \{2, 3\}$ then $Cay(\Gamma, S')$ is shown in Figure 11.

Example 5.2. $\Gamma = F_2 = \langle a, b\rangle$ the free group on 2 generators. Then $Cay(F_2, \{a, b\})$ is illustrated in Figure 12 below.

This is the universal cover of the figure-eight.

Different generating sets of a group may yield different Cayley graphs but they should be *related*. This notion is *quasi-isometry.*

Roughly speaking, a quasi-isometry is a map (not necessarily continuous) between metric spaces which distorts distance by a uniformly bounded amount, defined precisely as follows.

Definition 5.1. *Let X and Y be metric spaces. A map $f : X \to Y$ is a (K, C)-quasi-isometry, for $K \geq 1$ and $C \geq 0$ if*

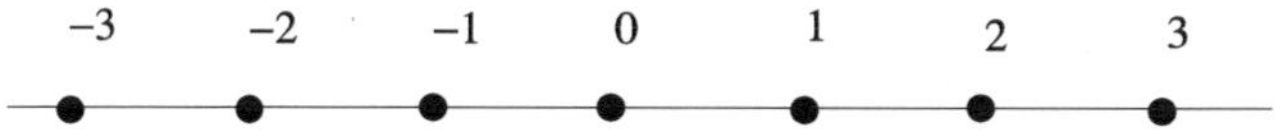

Figure 10. $Cay(\mathbb{Z}, \{1\})$

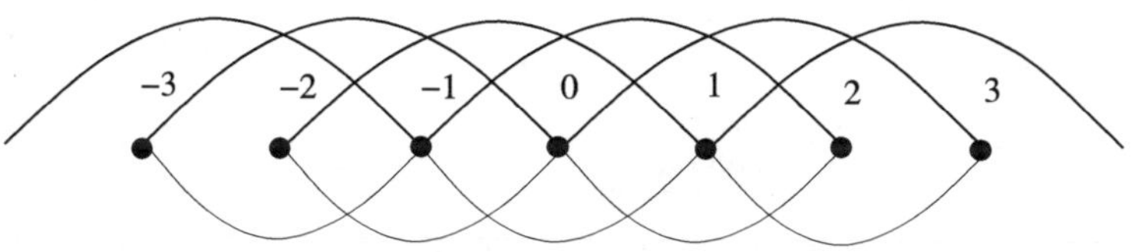

Figure 11. $Cay(\mathbb{Z}, \{2, 3\})$

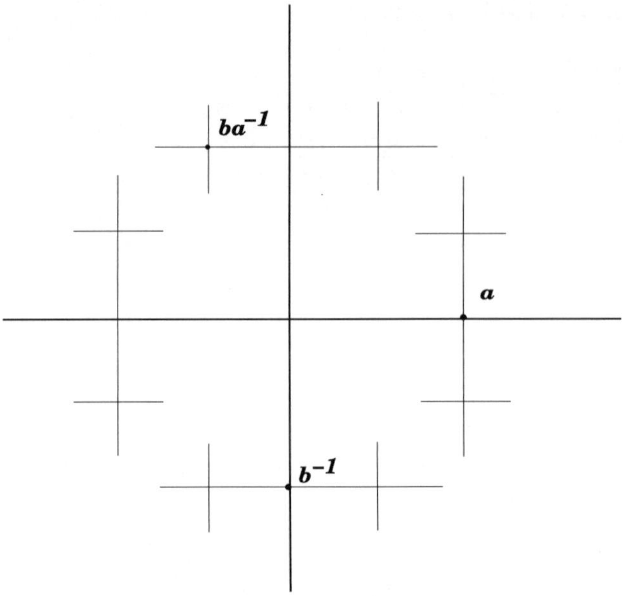

Figure 12. $Cay(F_2, \{a, b\})$ infinite snow-flake

(1) $\frac{1}{K}d_X(x_1, x_2) - C \leq d_Y(f(x_1), f(x_2)) \leq Kd_X(x_1, x_2) + C$ *for all* $x_1, x_2 \in X$.
(2) *For some constant* C', *we have* $Nhbd_{C'}(f(X)) = Y$.

(Note that the first condition is also referred as a quasi-isometric embedding.)

There is a notion of *coarse inverse* for a quasi-isometry; quasi-isometries f and g are coarse inverses if both compositions $f \circ g$ and $g \circ f$ are a bounded distance from the identity. Here, we say that X and Y are quasi-isometric. Without loss of generality we may assume that g and f share the same quasi-isometry constants.

Given a generating set S of a finitely generated group G, the inclusion $G \hookrightarrow Cay(G, S)$ is a quasi-isometry. Moreover, $Cay(G, S_1)$ and $Cay(G, S_2)$ are quasi-isometric for two generating sets S_1, S_2.

The set of all self quasi-isometries of a finitely generated group G is denoted by $QIMap(G)$. We form equivalence classes consisting of all quasi-isometries of G which differ by a uniformly bounded amount. This set of equivalence classes is called the *quasi-isometry group* of G and is denoted by $QI(G)$.

Recall that a group G acts (as homeomorphisms) *properly or properly discontinuously* on a locally compact space X if for every compact subspace K of X, the set $\{g \in G | gK \cap K \neq \emptyset\}$ is finite. A metric space X is *proper* if its closed balls of finite radius are compact; is *geodesic* if any two points x, y can be joined by at least one geodesic segment (i.e., $\exists$ an isometric embedding $\phi : [0, L] \to X$ such that $\phi(0) = x, \phi(L) = y$ and $L = d_X(x, y)$).

The following result, also known as the *Fundamental Theorem of Geometric Group Theory* (see p.87 of [36]), was known to Efremovich and to Svarc in the early 1950's and rediscovered by Milnor in 1968.

Theorem 5.1. *Let Γ be a finitely generated group acting co-compactly and properly on a proper geodesic metric space X. Then Γ is quasi-isometric to X. In fact, for any $x_0 \in X$, the map $\Gamma \to X$ given by $\gamma \mapsto \gamma x_0$ is a quasi-isometry.*

If G is a finitely generated group and X is a proper geodesic metric space, we define a *quasi-action* (see e.g. [51]) of G on X to be a map $\Psi : G \to QIMap(X)$ satisfying the following properties for some constants $K \geq 1$ and $C \geq 0$.

(1) For each $g \in G$, the element $\Psi(g)$ is a (K, C)-quasi-isometry of X.
(2) $\Psi(Id)$ is a uniformly bounded distance from the identity, that is, $\Psi(Id)$ is the identity in $QI(X)$.
(3) $\Psi(g)\Psi(h)$ is a uniformly bounded distance from $\Psi(gh)$, for all $g, h \in G$.

It is clear that a quasi-action induces a homomorphism from G into $QI(X)$. It is important to note that all quasi-isometries $\Psi(g)$ have the same quasi-isometry constants.

One of the major objectives of study in Gromov's program is the classification of finitely generated groups up to quasi-isometry (see e.g. [36]). Here are some basic facts about quasi-isometry.

(1) If Γ is finitely generated and Γ' is a subgroup of finite index in Γ then Γ' is finitely generated and Γ' is quasi-isometric to Γ ($\Gamma' \simeq_{q.i.} \Gamma$).
(2) Given a short exact sequence

$$1 \to \Gamma' \to \Gamma \to \Gamma'' \to 1$$

such that Γ' is finite and Γ'' is finitely generated, then Γ is finitely generated and $\Gamma \simeq_{q.i} \Gamma''$.
(3) If X is a compact Riemannian manifold with universal cover $\tilde{X}$ then $\Gamma = \pi_1(X, x_0)$ is quasi-isometric to $\tilde{X}$. In particular, if Σ_g denotes a closed surface of genus g then for $g > 1$, $\pi_1(\Sigma_g) \simeq_{q.i} \mathbb{H}^2$, the hyperbolic plane.
(4) Let G be a connected real Lie group and $\Gamma \subset G$ a uniform lattice (co-compact and discrete). Then Γ is finitely generated and $\Gamma \simeq_{q.i} G$.

Toward classifying groups up to quasi-isometry, one often considers quasi-isometry invariants. We say that a group property $\mathcal{P}$ is *geometric* if, whenever $G_1 \simeq_{q.i} G_2$, we have that G_1 has property $\mathcal{P}$ iff G_2 has property $\mathcal{P}$.

Here are some of the well-known geometric properties.

(i) Γ is finite; (ii) Γ is virtually $\mathbb{Z}$, i.e., it has a finite index subgroup isomorphic to $\mathbb{Z}$; (iii) $e(\Gamma)$, the number of ends of Γ; (iv) Γ is virtually free; (v) Γ has a free abelian subgroup of finite index; (vi) Γ is virtually nilpotent; (vii) Γ is finitely presented; Γ is Gromov hyperbolic; (viii) growth type of Γ (polynomial, intermediate, exponential); (ix) Γ is amenable (see [36] for definition; see also [70]).

For instance, being virtually solvable is *not* a geometric property. However, it is still unknown whether being virtually polycyclic is geometric.

Problem 6. Study the R_∞ property as a geometric property. In other words, determine the class of groups for which R_∞ is geometric.

The reason why we ask such questions comes from the following observation concerning $R(\varphi)$.

Consider a commutative diagram of short exact sequences where the vertical maps are group automorphisms.

$$\begin{array}{ccccccccc} 1 & \longrightarrow & A & \xrightarrow{i} & B & \xrightarrow{p} & C & \longrightarrow & 1 \\ & & \varphi' \downarrow & & \varphi \downarrow & & \overline{\varphi} \downarrow & & \\ 1 & \longrightarrow & A & \xrightarrow{i} & B & \xrightarrow{p} & C & \longrightarrow & 1 \end{array} \tag{5.1}$$

Then we obtain a short exact sequence of sets and corresponding functions $\hat{i}$ and $\hat{p}$:

$$\mathcal{R}(\varphi') \xrightarrow{\hat{i}} \mathcal{R}(\varphi) \xrightarrow{\hat{p}} \mathcal{R}(\overline{\varphi}), \tag{5.2}$$

where if $\bar{1}$ is the identity element in C, we have $\hat{i}(\mathcal{R}(\varphi')) = \hat{p}^{-1}([\bar{1}])$, and $\hat{p}$ is onto. The following result is straightforward.

Lemma 5.1. *Given the commutative diagram above,*

(1) *if* $R(\overline{\varphi}) = \infty$ *then* $R(\varphi) = \infty$,
(2) *if* C *is finite and* $R(\varphi') = \infty$ *then* $R(\varphi) = \infty$, *and*
(3) *if* A *and* C *are abelian,* $R(\overline{\varphi}) < \infty$ *with* $Fix(\overline{\varphi}) = 1$, *and* $R(\varphi') = \infty$ *then* $R(\varphi) = \infty$.

When $C \cong \mathbb{Z}$, *we have one of two situations:*

(4) *the map* $\overline{\varphi}$ *is the Identity, in which case* $R(\overline{\varphi}) = \infty$ *and hence* $R(\varphi) = \infty$, *or*
(5) $\overline{\varphi}(t) = t^{-1}$ *and* $R(\overline{\varphi}) = 2$. *In this case,* $R(\varphi) < \infty$ *iff* $R(\varphi') < \infty$ *and* $R(t \cdot \varphi') < \infty$, *in which case* $R(\varphi) = R(\varphi') + R(t \cdot \varphi')$.

Note that in case (2) when C is finite, $A \simeq_{q.i.} B$.

We will make use of the quasi-isometry rigidity and classification for the Baumslag-Solitar groups to show that R_∞ is geometric among these groups.

The solvable Baumslag-Solitar groups $BS(1, m)$ have presentation $\langle a, t | tat^{-1} = a^m \rangle$, and can be expressed as the group extension

$$1 \to \mathbb{Z}\left[\frac{1}{m}\right] \to BS(1, m) \to \mathbb{Z} \to 1.$$

The large-scale geometry of these groups was studied by Farb and Mosher [18,19], who describe the geometric model for these groups as the warped product of a tree and the real line. The metric on this complex is such that when a height function is put on the tree, and l is a line in the tree on which the height function is strictly increasing, $l \times \mathbb{R}$ is quasi-isometric to a hyperbolic plane.

First we consider a solvable generalization of $BS(1, m)$, which we define in two stages. The initial generalization arises in a geometric context, but easily extends to a strictly algebraic generalization. When we view the group $PSL_2(\mathbb{Z}[\frac{1}{p}])$, for p a prime, as a subgroup of $PSL_2(\mathbb{R}) \times PSL_2(\mathbb{Q}_p)$ under the diagonal embedding, we get a natural action of $PSL_2(\mathbb{Z}[\frac{1}{p}])$ on $\mathbb{H}^2 \times T$ where T is the Bruhat-Tits tree associated to $PSL_2(\mathbb{Q}_p)$. Here $\mathbb{Q}_p$ denotes the p-adic rationals. Under this action, the stabilizer of a point at infinity is exactly $BS(1, p^2)$, which is commensurable with $BS(1, p)$.

We can analogously consider the group $PSL_2(\mathbb{Z}[\frac{1}{n}])$, where n is any positive integer, which acts on the product of $\mathbb{H}^2$ with the product of the Bruhat-Tits trees for $PSL_2(\mathbb{Q}_{p_i})$. Here, p_i are the prime factors of n. The stabilizer of a point at infinity under this action is the upper triangular subgroup of $PSL_2(\mathbb{Z}[\frac{1}{n}])$, which we denote Γ_n. This group is also defined by the group extension $1 \to \mathbb{Z}[\frac{1}{n}] \to \Gamma_n \to \mathbb{Z}^k \to 1$. We use the following presentation of Γ_n:

$$\Gamma_n = \langle a, t_1, t_2, \ldots, t_k | t_i t_j = t_j t_i,\ i \neq j,\ t_i^{-1} a t_i = a^{p_i} \rangle.$$

Note that when $k = 1$ this is exactly $BS(1, p)$. The geometric model for these groups is topologically the product of $\mathbb{R}$ with the product of trees on which the group acts.

The groups Γ_n generalize immediately to a larger class of groups $\Gamma(S)$ which do not arise in the same geometric context, but have similar algebraic properties to Γ_n. Let $S = \{n_1, n_2, \ldots, n_k\}$ be a set of pairwise relatively prime positive integers. We define $\Gamma(S)$ to be the group

$$\Gamma(S) = \langle a, t_1, t_2, \ldots, t_k | t_i t_j = t_j t_i,\ i \neq j,\ t_i^{-1} a t_i = a^{n_i} \rangle$$

and we use this presentation throughout the paper. We will use the following definition of $\Gamma(S)$ as a group extension below:

$$1 \to A \to \Gamma(S) \xrightarrow{\pi} \mathbb{Z}^k \to 1,$$

where the map $\pi : \Gamma(S) \to \mathbb{Z}^k$ is a "component-wise height function" with the i-th coordinate in $\pi(g)$ being the sum of the exponents of the generator t_i in a word representing $g \in \Gamma(S)$. As with $BS(1, m)$ and Γ_n, the kernel A of π is the normal closure of the subgroup generated by the element a in the presentation above. These groups have geometric models analogous to those for Γ_n.

In 2007, J. Taback and the author [56] proved that the groups $\Gamma(S)$ have R_∞. For each $\psi \in QI(\Gamma(S))$, the geometric idea is to find a copy $\mathbb{Z}^k_*$ of $\mathbb{Z}^k$ inside the geometric model on which $\Gamma(S)$ acts such that ψ fixes $\mathbb{Z}^k_*$ pointwise. To prove that any group quasi-isometric to $\Gamma(S)$ has the R_∞ property, we make use of a quasi-action of the group on $QI(Map(X))$, where the geometric model X is a product of the real line with a product of trees.

Recently, G. Levitt [47] studied the automorphism group of a generalized Baumslag-Solitar group, that is, the fundamental group of a graph of groups where the vertex/edge groups are $\mathbb{Z}$'s. He showed that non-elementary GBS groups have the R_∞ property.

J. Taback and the author [57] also extended Levitt's result by showing the following result.

Theorem 5.2. *Let G be a finitely generated group quasi-isometric to a non-elementary generalized Baumslag-Solitar group. Then G has property R_∞.*

The idea of the proof falls into three parts.

First, K. Whyte [64] showed that a GBS group Γ is either:

(i) $BS(1, n)$ for some $n > 1$;
(ii) is virtually $F_n \times \mathbb{Z}$;
(iii) quasi-isometric to $BS(2, 3)$.

Case i. use Farb-Mosher [18,19] quasi-isometry classification of $BS(1, n)$ and show that there is a copy of $\mathbb{Z}$ inside Γ that is pointwise fixed. (Also true for more general situations.)

Case ii. G is also virtually $F_n \times \mathbb{Z}$. Using a result of Kleiner-Leeb [46], we have

$$1 \to H \to G \to L \to 1$$

such that H is virtually $\mathbb{Z}$ and characteristic; L is Gromov hyperbolic. Now $\varphi \in \mathrm{Aut}(G)$ induces $\bar{\varphi} \in \mathrm{Aut}(L)$ and $R(\bar{\varphi}) = \infty$ so $R(\varphi) = \infty$,

Case iii. $G \simeq_{q.i.} BS(2, 3)$. It follows from a result of K. Whyte [64] that G is π_1 of a graph of virtual $\mathbb{Z}$'s which is neither commensurable to $F_n \times \mathbb{Z}$ nor virtually solvable. Following a similar argument as Levitt [47] by considering the elliptic elements of G, one can show that G has R_∞.

Remark 5.1. The difference (from Levitt's work) is that we have a graph of **virtual** $\mathbb{Z}$'s and we have a **quasi**-action on a Cayley complex instead of a genuine action.

We end this paper by using geometric methods to study the R_∞ property for lamplighter groups $G \wr \mathbb{Z}$, where $\wr$ denotes the restricted wreath product.

We begin with a discussion of the geometry of the lamplighter groups $L_n = \mathbb{Z}_n \wr \mathbb{Z}$ and later address more general groups $G \wr \mathbb{Z}$ where G is any finitely generated group.

The standard presentation of L_n is $\langle a, t | a^n = 1,\ [a^{t^i}, a^{t^j}] = 1\rangle$ where x^y denotes the conjugate yxy^{-1}. Equivalently, we recall that a wreath product is simply a certain semidirect product, and write L_n as $(\bigoplus_{i=-\infty}^{\infty} A_i) \rtimes \mathbb{Z}$, with each $A_i \cong Z_n$, fitting into the split short exact sequence

$$0 \to \bigoplus_{i=-\infty}^{\infty} A_i \to L_n \to \mathbb{Z} \to 0$$

where the generator of $\mathbb{Z}$ is taken to be t from the presentation given above.

The "lamplighter" picture of elements of this group is the following. Take a bi-infinite string of light bulbs placed at integer points along the real line, each of which has n states corresponding to the powers of the generator a of $\mathbb{Z}_n$, and a cursor, or lamplighter, indicating the particular bulb under consideration. The action of the group on this picture is such that t moves the cursor one position to the right, and a changes the state of the current bulb under consideration. Thus each element of L_n can be interpreted as a series of instructions to illuminate a finite collection of light bulbs to some allowable states, leaving the lamplighter at a fixed integer.

Understanding group elements via this picture, the generator $a_j = t^j a t^{-j}$ of A_j in $\bigoplus_{i=-\infty}^{\infty} A_i$ has a single bulb in position i illuminated to state a, and the cursor at the origin of $\mathbb{Z}$. This leads to two possible normal forms for $g \in L_n$, as described in [11], separating the word into segments corresponding to bulbs indexed by negative and non-negative integers:

$$rf(g) = a_{i_1}^{e_1} a_{i_2}^{e_2} \dots a_{i_k}^{e_k} a_{-j_1}^{f_1} a_{-j_2}^{f_2} \dots a_{-j_l}^{f_l} t^m$$

or

$$lf(g) = a_{-j_1}^{f_1} a_{-j_2}^{f_2} \dots a_{-j_l}^{f_l} a_{i_1}^{e_i} a_{i_2}^{e_i} \dots a_{i_k}^{e_k} t^m$$

with $i_k > \cdots i_2 > i_1 \geq 0$ and $j_l > \cdots j_2 > j_1 > 0$ and e_i, f_j are in the range $\{-h, \dots, h\}$, where h is the integer part of $\frac{n}{2}$. When n is even, we omit a^{-h} to ensure uniqueness, since $a^h = a^{-h}$ in $\mathbb{Z}_{2h}$. If we consider the normal form as a series of instructions for acting on the lamplighter picture to create these group elements, then the normal form $rf(g)$ first illuminates bulbs at or to the right of the origin, and $lf(g)$ first illuminates the bulbs to the left of the origin. It is proven in [11] that the word length of $g \in L_n$ with respect to the finite generating set $\{a, t\}$, using the notation of either normal form given above, is

$$\Sigma e_i + \Sigma f_j + \min\{2j_l + i_k + |m - i_k|, 2i_k + j_l + |m + j_l|\}.$$

We note that the lamplighter picture of L_n is not the Cayley graph of the group; to obtain a useful and understandable Cayley graph for this group we introduce Diestel-Leader graphs below. However, to use a Diestel-Leader graph (see e.g. [65]) as the Cayley graph of L_n, we must consider the generating set $\{t, ta, ta^2, \dots, ta^{n-1}\}$ for L_n.

We now describe explicitly the Cayley graphs for the lamplighter group L_m with respect to one particular generating set. This Cayley graph is an example of a Diestel-Leader graph, which we define in full generality as follows.

For positive integers $m \leq n$, the Diestel-Leader graph $DL(m, n)$ is a subset of the product of the regular trees of valence $m + 1$ and $n + 1$, which we denote respectively as T_1 and T_2. We orient these trees so that each vertex has m (resp. n) outgoing edges. By fixing a basepoint, we can use this orientation to define a height function $h_i : T_i \to \mathbb{Z}$. The Diestel-Leader graph $DL(m, n)$ is defined to be the subset of $T_1 \times T_2$ for which $h_1 + h_2 = 0$. This definition lends itself to the following pictorial representation. Each tree has a distinguished point at infinity which determines the height function. If we place these points for the two trees at opposite ends of a page, the Diestel-Leader graph can be seen as those pairs of points, one from each tree, lying on the same horizontal line. See Figure 13 for the graph $DL(3, 3)$.

Using the recent quasi-isometry classification of lamplighter groups due to Eskin-Fisher-Whyte [14,15] which says that every quasi-isometry is a uniformly bounded distance from a product map on $T_1 \times T_2$, J. Taback and the author [58] use the geometry to prove that $\mathbb{Z}_n \wr \mathbb{Z}$ has R_∞ iff $2|n$ or $3|n$, a result first proved by D. Gonçalves and the author in [32] using algebraic methods.

Problem 7. Determine the R_∞ property for other lamplighter groups $G \wr \mathbb{Z}$, for non-abelian finite groups G.

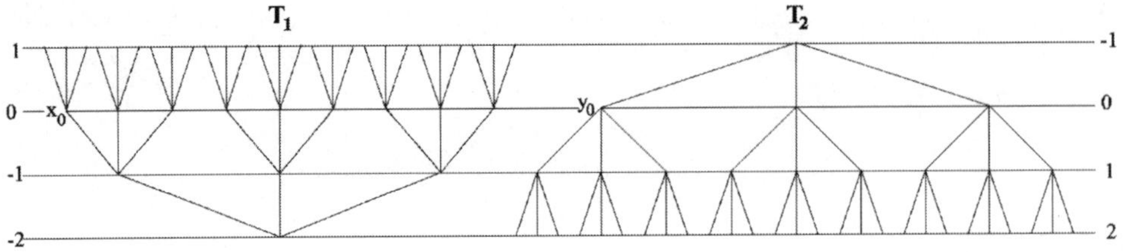

Figure 13. Part of the Cayley graph $DL(3, 3)$ of L_3 with the point $Id = (x_0, y_0)$ labeled. The integers on the edges of the diagram represent the height in each tree.

References

[1] D. Anosov, The Nielsen number of maps of nil-manifolds, *Russian Math. Surveys*, **40** (1985) 149–150.

[2] A. J. Berrick, I. Chatterji and G. Mislin, Homotopy idempotents on manifolds and Bass' conjectures, *Geom. Topol. Monogr.*, **10** (2006) 41–62.

[3] O. Bogopolski, A. Martino, O. Maslakova and E. Ventura, The conjugacy problem is solvable in free-by-cyclic groups, *Bull. London Math. Soc.*, **38** (2006) no. 5, 787–794.

[4] P. Brinkmann, Dynamics of free group automorphisms, preprint (2003).

[5] Robert F. Brown, "The Lefschetz Fixed Point Theorem," Scott-Foresman, Glenview, Illinois, (1971).

[6] Robert F. Brown, Fixed point theory, *History of topology*, 271–299, North-Holland, Amsterdam, (1999).

[7] Robert F. Brown, The Nielsen number of a fibre map, *Ann. Math.*, **85** (1967) 483–493.

[8] Robert F. Brown and E. Fadell, Corrections to "The Nielsen number of a fibre map," *Ann. Math.*, **95** (1972) 365–367.

[9] Handbook of topological fixed point theory, Edited by Robert F. Brown, Massimo Furi, Lech Górniewicz and Boju Jiang, Springer, Dordrecht (2005).

[10] L. Charlap, "Bieberbach Groups and Flat Manifolds," Springer, New York (1986).

[11] S. Cleary and J. Taback, Dead end words in lamplighter groups and other wreath products, *Quarterly Journal of Mathematics*, **56** (2005) no. 2, 165–178.

[12] K. Dekimpe, B. DeRock and P. Penninckx, The Anosov theorem for infra-nil-manifolds, preprint (2007).

[13] J. Dyer, A nilpotent Lie algebra with nilpotent automorphism group, *Bull. Amer. Math. Soc.*, **76** (1970) 52–60.

[14] A. Eskin, D. Fisher and K. Whyte, Coarse differentiation of quasi-isometries I: spaces not quasi-isometric to Cayley graphs, preprint (2007).

[15] A. Eskin, D. Fisher and K. Whyte, Coarse differentiation of quasi-isometries II: Rigidity for Sol and Lamplighter groups, preprint (2007).

[16] E. Fadell and S. Husseini, On a theorem of Anosov on Nielsen numbers for Nil-manifolds, Nonlinear functional analysis and its applications (Maratea, 1985) NATO, Advanced Science Institutes Series C: Mathematical and Physical Science 173 (Kluwer, Dordrecht, 1986) 47–53.

[17] E. Fadell and S. Husseini, The Nielsen number on surfaces, Topological methods in nonlinear functional analysis (Toronto, Ont., 1982) 59–98, *Contemp. Math.*, **21**, *Amer. Math. Soc.*, Providence, RI (1983).

[18] B. Farb and L. Mosher (appendix by D. Cooper), A rigidity theorem for the solvable Baumslag-Solitar groups, *Inventiones Math.*, **131** (1998) no. 2, 419–451.

[19] B. Farb and L. Mosher Quasi-isometric rigidity for the solvable Baumslag-Solitar groups, II, *Inventiones Math.*, **137** (1999) no. 3, 273–296.

[20] A. L. Fel'shtyn, The Reidemeister number of any automorphism of a Gromov hyperbolic group is infinite, *Zapiski Nauchnych Seminarov POMI*, **279** (2001) 229–241.

[21] A. L. Fel'shtyn and D. Gonçalves, Twisted conjugacy classes of automorphisms of Baumslag-Solitar groups, *Algebra and Discrete Math.*, (2006) no. 3, 36–48.

[22] A. L. Fel'shtyn and D. Gonçalves, Twisted conjugacy classes in symplectic groups, Mapping class groups and Braid groups (including an Appendix written with Francois Dahmani), preprint (2007).

[23] A. L. Fel'shtyn, D. Gonçalves and P. Wong, Twisted conjugacy classes of automorphisms of polyfree groups, preprint (2008).

[24] A. L. Fel'shtyn and R. Hill, The Reidemeister zeta function with applications to Nielsen theory and a connection with Reidemeister torsion, *K-theory*, **8** (1994) no. 4, 367–393.

[25] A. L. Fel'shtyn, R. Hill and P. Wong, Reidemeister numbers of equivariant maps, *Top. Appl.*, **67** (1995) 119–131.

[26] A. Fel'shtyn and E. Troitsky, Twisted Burnside-Frobenius theory for discrete groups, *J. Reine Angew. Math.*, **613** (2007) 193–210.

[27] W. Franz, Abbildungsklassen und Fixpunktklassen dreidimensionaler Linsenräume, (German) *J. Reine Angew. Math.*, **185** (1943) 65–77.

[28] D. L. Gonçalves, The coincidence Reidemeister classes of maps on nilmanifolds, *Topol. Methods Nonlinear Anal.*, **12** (1998) no. 2, 375–386.

[29] D. Gonçalves and P. Wong, Twisted conjugacy classes in exponential growth groups, *Bull. London Math. Soc.*, **35** (2003) 261–268.

[30] D. Gonçalves and P. Wong, Homogeneous spaces in coincidence theory, II, *Forum Math.*, **17** (2005) 297–313.

[31] D. Gonçalves and P. Wong, Homogeneous spaces in coincidence theory, In: P. Schweitzer, ed., Conference proceedings of the Tenth Brazilian Topology Meeting (São Carlos 1996), *Mat. Contemp.*, **13** (1997) 143–158.

[32] D. Gonçalves and P. Wong, Twisted conjugacy classes in wreath products, *Internat. J. Alg. Comput.*, **16** (2006) 875–886.

[33] D. Gonçalves and P. Wong, Twisted conjugacy classes in nilpotent groups, *J. reine angew. Math.*, **633** (2009) 11–27.

[34] D. Gottlieb, A certain subgroup of the fundamental group, *Amer. J. Math.*, **87** (1965) 840–856.

[35] Andrzej Granas and James Dugundji, "Fixed point theory," Springer Monographs in Mathematics, Springer-Verlag, New York (2003).

[36] P. de la Harpe, Topics in geometric group theory, Chicago Lectures in Mathematics University of Chicago Press, Chicago, IL (2000) vi+310 pp.

[37] P. Heath, E. Keppelmann and P. Wong, Addition formulae for Nielsen numbers and for Nielsen type numbers of fiber-preserving maps, *Top. Appl.*, **67** (1995) 133–157.

[38] G. Higman, B. Neumann and H. Neumann, Embedding theorems for groups, *J. London Math. Soc.*, **24** (1949) 247–254.

[39] S. Ivanov, On some finiteness conditions in semigroup and group theory, *Semigroup Forum*, **48** (1994) no. 1, 28–36.

[40] Boju Jiang, "Lectures on Nielsen Fixed Point Theory," *Contemp. Math.*, **14** *Amer. Math. Soc., Providence* (1983).

[41] B. Jiang, S. Wang and Y. Wu, Homeomorphisms of 3-manifolds and the realization of Nielsen number, *Comm. Anal. Geom.*, **9** (2001) 825–877.

[42] B. Jiang and J. Guo, Fixed points of surface diffeomorphisms, *Pacific J. Math.*, **160** (1993) 67–89.

[43] M. Kelly, The Nielsen number as an isotopy invariant, *Topology Appl.*, **62** (1995) no. 2, 127–143.

[44] Tsai-han Kiang, "The Theory of Fixed Point Classes," Springer, New York (1987).

[45] S. Kim, Computation of Nielsen numbers for maps of compact surfaces with boundary, *J. Pure Appl. Algebra*, **208** (2007) no. 2, 467–479.
[46] B. Kleiner and B. Leeb, Groups quasi-isometric to symmetric spaces, *Comm. Anal. Geom.*, **9** (2001) no. 2, 239–260.
[47] G. Levitt, On the automorphism group of generalized Baumslag-Solitar groups, *Geom. Topol. Geom. Topol.*, **11** (2007) 473–515.
[48] G. Levitt and M. Lustig, Most automorphisms of a hyperbolic group have very simple dynamics, *Ann. Scient. Éc. Norm. Sup.*, **33** (2000) 507–517.
[49] W. Magnus, A. Karrass and D. Solitar, Combinatorial Group Theory. New York: Interscience Publishers (1966).
[50] O. Maslakova, The fixed point group of a free group automorphism, *Algebra Logika*, **42** (2003) 422–472 (in Russian), *Algebra Logic*, **42** (2003) 237–265 (in English).
[51] L. Mosher, M. Sageev and K. Whyte, Quasi-actions on trees I, *Ann. of Math.*, (2) **158** (2003) 115–164.
[52] D. Osin, Small cancellations over relatively hyperbolic groups and embedding theorems, preprint (2004).
[53] Z. Sela, Structure and rigidity in (Gromov) hyperbolic groups and discrete groups in rank 1 Lie groups. II. *Geom. Funct. Anal.*, **7** (1997) no. 3, 561–593.
[54] P. Staecker, The Reidemeister trace: computation by nilpotentization and extension to coincidence theory, Ph.D. dissertation, UCLA (2005).
[55] J. Taback and K. Whyte, Twisted conjugacy and group actions, preprint (2005).
[56] J. Taback and P. Wong, Twisted conjugacy and quasi-isometry invariance for generalized solvable Baumslag-Solitar groups, *J. London Math. Soc.* (2), **75** (2007) 705–717.
[57] J. Taback and P. Wong, A note on twisted conjugacy and generalized Baumslag-Solitar groups, preprint (2006).
[58] J. Taback and P. Wong, The geometry of twisted conjugacy classes in wreath products, preprint (2008).
[59] K. Varadarajan, these proceedings.
[60] James Vick, "Homology Theory," Academic Press, New York (1973).
[61] J. Wagner, An algorithm for calculating the Nielsen number on surfaces with boundary, *Trans. Amer. Math. Soc.*, **351** (1999) no. 1, 41–62.
[62] F. Wecken, Fixpunktklassen, II, Homotopieinvarianten der Fixpunkttheorie (German), *Math. Ann.*, **118** (1941) 216–234.
[63] F. Wecken, Fixpunktklassen, III, Mindestzahlen von Fixpunkten (German), *Math. Ann.*, **118** (1942) 544–577.
[64] K. Whyte, The large scale geometry of the higher Baumslag-Solitar groups, *Geom. Funct. Anal.*, **11** (2001) no. 6, 1327–1343.
[65] W. Woess, Lamplighters, Diestel-Leader graphs, random walks, and harmonic functions, *Combinatorics, Probability & Computing*, **14** (2005) 415–433.
[66] P. Wong, Fixed point theory for homogeneous spaces—a brief survey. Handbook of topological fixed point theory, 265–283, Springer, Dordrecht (2005).
[67] P. Wong, Reidemeister number, Hirsch rank, coincidences on polycyclic groups and solvmanifolds, *J. reine angew. Math.*, **524** (2000) 185–204.
[68] K. Wortman, A finitely presented solvable group with small quasi-isometry group, *Michigan Math. J.*, **55** (2007) no. 1, 3–24.
[69] P. Yi, An algorithm for computing the Nielsen number of maps on the pants surface, Ph.D. dissertation, UCLA (2003).
[70] P. Zvengrowski, these proceedings.

Perspectives in Geometry and Topology No. 11, 2010, pp. 133–170.

Mini-Course: An Introduction to ℓ_2-Homology

Peter Zvengrowski

Department of Mathematics and Statistics, University of Calgary, Calgary, Alberta T2N 1N4, Canada
e-mail: zvengrow@ucalgary.ca

Dedicated to the Memory of Beno Eckmann, 1917–2008, and of Heiner Zieschang, 1936–2004

Contents

Notation

Here are some frequently used notations in these notes and in the mini-course.

$\subset$	proper subset
$\subseteq$	subset
e	the neutral (identity) element of a group G
$X^{(n)}$	the n-skeleton of a CW-complex
aa	almost all (all but a finite number)
st	such that
TFAE	the following are equivalent
wlog	without loss of generality
wrt	with respect to
$\boxplus$	orthogonal internal direct sum decomposition
$\approx$	isomorphic
$\cong$	homeomorphic

1. Introduction

The inspiration for this mini-course comes from similar lectures given by Beno Eckmann during one of his visits to western Canada, and the author's subsequent attempts to understand this fascinating subject. In 2000 a substantial paper of Eckmann's appeared in the Israel J. Math [5], based on the notes (by Guido Mislin) from a mini-course he gave in 1997 at the Mathematical Research Institute, ETH Zürich. The present notes are based on these notes of Eckmann, with very little, if any, claim to originality. An introductory chapter (Chapter II) on basic Hilbert space theory has been added, since the subsequent material is completely based on this. Most but not all of the Eckmann paper is covered, however it is the author's hope that the present mini-course will give a fairly thorough introduction to the basics of the subject, so that anyone attending the mini-course should have no great difficulty in reading the remainder of [5] or other literature involving ℓ_2-homology, notably the book of Lück [16].

The subject originated with Atiyah's ideas (cf. [1]), in 1976, of applying Hilbert space technique to algebraic topology to obtain refined (ℓ_2) invariants of (generally infinite) cell complexes. It has been used by many authors since then and become an increasingly important tool, notably the ideas of ℓ_2-homology (ℓ_2-cohomology) and the associated ℓ_2-Betti numbers β_i. Applications of the theory have been made to low dimensional topology, K-theory, group theory (in particular the study of amenable groups), ring theory, geometry, and other areas.

We shall now attempt to give some intuitive feeling for the subject and its applications by means of three examples, the first two quite elementary.

Example 1.1.1. A nice place to start any topological discussion is Euler's famous formula $V - E + F = 2$. In more modern language, for S^2 the 2-sphere, one says $\chi(S^2) = 2 = \alpha_0 - \alpha_1 + \alpha_2$, where χ is the Euler characteristic and α_i the number of i-cells. One also has the homology groups $H_0(S^2) \approx \mathbb{Z}$, $H_2(S^2) \approx \mathbb{Z}$, and otherwise

$H_i(S^2) = 0$. The rank of the finitely generated abelian group H_i is called the ith Betti number b_i of the space. It can also be defined by taking homology with coefficients in $\mathbb{Q}$ or $\mathbb{R}$ (resp. rational or real homology), and taking the dimension of the resulting vector space. Thus $b_0(S^2) = \dim_{\mathbb{R}} H_0(S^2; \mathbb{R}) = 1$, $b_2(S^2) = \dim_{\mathbb{R}} H_2(S^2; \mathbb{R}) = 1$, and $b_i(S^2) = 0$ otherwise. This illustrates a well known theorem of algebraic topology,

$$\chi(X) = \sum_{i \geq 0} (-1)^i \alpha_i(X) = \sum_{i \geq 0} (-1)^i b_i(X),$$

for X a finite CW-complex. In the present case $\chi(S^2) = 2 = b_0 - b_1 + b_2 = 1 - 0 + 1$.

For S^1, $\chi(S^1) = 0$, $b_0(S^1) = b_1(S^1) = 1$ and $b_i(S^1) = 0$ otherwise. So here the theorem takes the form $\chi(S^1) = 0 = b_0 - b_1 = 1 - 1$.

Example 1.1.2. Now consider the familiar covering projection $p : \mathbb{R} \twoheadrightarrow S^1$, $p(t) = \exp(2\pi i t)$. Here S^1 is the unit circle in $\mathbb{C}$, with base point 1, and $\mathbb{R}$ has base point 0. Let $S^1 = e^0 \cup e^1$, a cellular decomposition with one 0-cell $e^0 = \{1\}$ and one 1-cell e^1. The corresponding cells of $\mathbb{R}$ can be written $\bar{e}^0_j = \{j\} \subset \mathbb{R}$, $j \in \mathbb{Z}$, and $\bar{e}^1_j = (j, j+1) \subset \mathbb{R}$. The fundamental group $\pi_1(S^1, 1) = \mathbb{Z}$ acts freely on $\mathbb{R}$ by integral translations, this action is cellular (permutes the cells) and free (only the identity $0 \in \mathbb{Z}$ fixes any cell). This is an example of a regular covering, i.e. $p_* \pi_1(\mathbb{R}, 0) = \pi_*\{e\} = \{e\}$, the trivial group, is a normal subgroup of $\pi_1(S^1, 1)$. It is also an example of a cocompact group action by $G = \mathbb{Z}$ on $Y = \mathbb{R}$, namely $X = Y/G = S^1$ is compact.

Now trying to define $\alpha_0, \alpha_1, \ldots$ here would be futile, since $\alpha_0(Y), \alpha_1(Y)$ are infinite. To define (real) homology of Y one starts with the chains $K_i(Y; \mathbb{R})$, the real vector space with basis the i-cells of Y. Thus an element of, say $K_1(Y; \mathbb{R})$ is a sum $\sum_{j \in \mathbb{Z}} r_j \bar{e}^1_j$, with $r_j \in \mathbb{R}$, almost all r_j equal 0. We could consider a more "global" chain by allowing infinite sums, i.e. remove the condition almost all $r_j = 0$. But this will create other problems, e.g. the formula for the boundary map may well have divergent sums. This can be overcome by considering the ℓ_2-chains, where we impose the condition of square summability $\sum_{j \in \mathbb{Z}} r_j^2 < \infty$.

A chain complex of Hilbert spaces is thus obtained, with extra structure as G-modules arising from the action of G. The resulting homology groups, modulo a few technical details, are the ℓ_2-homology groups. They are generally (countably) infinite dimensional Hilbert spaces, but they have another equivariant type dimension, called the von Neumann dimension β_i. It turns out that β_i is a non-negative real number, which we call the ith ℓ_2-Betti number, and the Euler characteristic can be computed in terms of these. In the present example it turns out that $\beta_i(S^1) := \beta_i(\mathbb{R}, \mathbb{Z}) = 0$, $i \geq 0$, so one obtains the (not very exciting) formula

$$\chi(S^1) = 0 = \sum_{i \geq 0} (-1)^i \cdot 0.$$

Example 1.1.3. In this example we illustrate an application to algebra. Let G be a finitely presented group (finitely many generators and finitely many relations).

In general many different presentations are possible, the deficiency of any given presentation P is the number of generators g_P less the number of relations r_P, i.e. $g_P - r_P$. Since r_P can be increased at will without changing G, say by simply repeating the same relation, $g_P - r_P$ has no lower bound. However, by considering the abelianization G_{ab} of G it is easily seen (cf. 7.2.1) that $g_P - r_P \leq \text{rank}(G_{ab}) := b_1(G)$, the first Betti number of G (or equivalently of the Eilenberg-MacLane space $K(G, 1)$). Thus the deficiency of any presentation P has an upper bound, and we define $\text{def}(G) := \max\{g_P - r_P : P \text{ is finite presentation of } G\}$. From the above, def $(G) \leq b_1(G)$. The difficult part of finding def (G) will generally be finding as sharp an upper bound as possible. One can then hope to find a presentation P achieving this bound. Thus, theorems of the following type can be quite useful.

Theorem. *def* $(G) \leq 1 + \beta_1(G)$ *(cf. §7.2).*

For example, for the free group F_n on n letters, we shall show def $(F_n) = n$. Similarly, for the fundamental group σ_g of an orientable surface of genus g, we shall show def $(\sigma_g) = 2g - 1$. For F_n the upper bound can be obtained using the above inequalities and b_1 or β_1. However, for σ_g, the above inequality with b_1 will not suffice to give the required upper bound, whereas the inequality with β_1 does suffice (a refined inequality 7.2.2 involving b_1, b_2 will also suffice).

Remark. In Example 1.1.2, although the number of cells α_0, α_1 for $Y = \mathbb{R}$ are infinite, one could argue that homologically $\mathbb{R}$ is quite simple, with $b_0 = 1$ and $b_j = 0, j > 0$. However, these (ordinary) Betti numbers are unrelated to the G action, and in slightly more complicated examples, the ordinary Betti numbers of Y, just like the α_i, can also be infinite.

The author wishes to heartily thank XIV Encontro Brasileiro de Topologia (Campinas, Brazil) for the opportunity to present this material as a mini-curso in July 2004, as well as the 25th Winter School on Geometry and Physics (January 2005, Srní, Czech Republic) and the Indian Institute of Technology Bombay (August 2007) for the opportunity to also present the mini-course there. In every case discussions with the participants of these meetings led to useful improvements in the manuscript, in particular the author is grateful to Pierre Cartier, Anthony Lau, Peter Michor, Parameswaran Sankaran, and especially to Jonathan Hillman, for their suggestions. Further thanks are due to Dr. Satya Prakash Tripathi, Rui Wang, and Lei Xiong for their most generous help in the typing of these notes. The mini-course is envisaged to take four to six hour lectures, with additional material added at the end of some chapters, as appendices, that can be omitted from the lectures, covered if time permits, or left for supplementary reading.

It is with great honour and humility that this mini-course is dedicated to the memory of both Professor Beno Eckmann of the Eidgenössische Technische Hochschule, Zürich, and Professor Heiner Zieschang of Ruhr-Universität Bochum. Each was a great friend, mathematician and inspiration to all who knew him. As already mentioned in this Introduction, the present set of notes owes its very existence to Eckmann's survey paper [5] as well as the excellent lectures he gave on this subject, to which he devoted much of his energy since about 1990.

2. Hilbert space, a brief review

This chapter gives a quick review of basic material on Hilbert space. It contains quite standard material (cf. [8,19]), apart perhaps from Lemma 2.3.10, and can be safely omitted by analysts or anyone who has had a course on Hilbert space in the not too distant past.

2.1 *Introduction to Hilbert space*

Definition 2.1.1. *An $\mathbb{R}$-vector space M together with a map $\langle\ ,\ \rangle : M \times M \to \mathbb{R}$ is an $\mathbb{R}$-inner product space if $\langle\ ,\ \rangle$ is bilinear, symmetric, and strictly positive.*

We shall only need $\mathbb{R}$-vector spaces in these notes. The corresponding definition for $\mathbb{C}$-vector spaces, incorporating conjugation is standard.

Definition 2.1.2. *For $x \in M$, $\|x\| = \langle x, x\rangle^{1/2}$ is the associated norm.*

Definition 2.1.3. *For $x, y \in M$, $x \perp y$ if and only if $\langle x, y\rangle = 0$. We say x, y are orthogonal.*

Proposition 2.1.4. *The following are equivalent:*

(a) $x = 0$,
(b) $\langle x, y\rangle = 0$ *for all* $y \in M$,
(c) $x \perp y$ *for all* $y \in M$,
(d) $\|x\| = 0$.

Polarization identity 2.1.5. $\langle x, y\rangle = \frac{1}{4}(\|x + y\|^2 - \|x - y\|^2)$.

Parallelogram law 2.1.6. $\|x + y\|^2 + \|x - y\|^2 = 2\|x\|^2 + 2\|y\|^2$.

Corollary 2.1.7 (Pythagorean Theorem). *For $x, y \in M$,*

$$x \perp y \quad \text{iff} \quad \|x \pm y\|^2 = \|x\|^2 + \|y\|^2.$$

Proof. Since $x \perp y$, 2.1.5 shows that $\|x + y\| = \|x - y\|$, then use 2.1.6. □

Schwarz inequality 2.1.8. For $x, y \in M$, $|\langle x, y\rangle| \le \|x\| \cdot \|y\|$, with equality if and only if x, y are linearly dependent.

Proof. For any $s, t \in \mathbb{R}$, one has

$$0 \le \|sx + ty\|^2 = [s\ \ t] \cdot A \cdot \begin{bmatrix} s \\ t \end{bmatrix},$$

where

$$A = \begin{bmatrix} \|x\|^2 & \langle x, y\rangle \\ \langle y, x\rangle & \|y\|^2 \end{bmatrix}$$

(called the Gram matrix of x, y) is symmetric, positive semi-definite. So A has real eigenvalues $\lambda_1, \lambda_2 \ge 0$, whence $\det A = \lambda_1 \cdot \lambda_2 \ge 0$. Since $\det A = \|x\|^2\|y\|^2 - \langle x, y\rangle^2$, this gives the Schwarz inequality. The case of equality is easily verified. □

Note that the same proof shows that the Gram matrix $A = [\langle x_i, x_j\rangle]$ of any n vectors $x_1, \ldots, x_n$ is symmetric positive semi-definite.

Using the Schwarz inequality (in 2.1.9(c) below) we see that $\|x\|$ satisfies the axioms for a norm.

Proposition 2.1.9. *The axioms for a norm hold for* $\|x\|$, *namely*

(a) $\|x\| \geq 0, \quad \|x\| = 0 \;$ iff $\; x = 0,$
(b) $\|rx\| = |r| \cdot \|x\|, \quad r \in \mathbb{R},$
(c) *Triangle inequality:* $\|x + y\| \leq \|x\| + \|y\|.$

Proof of (c). $\|x + y\|^2 = \|x\|^2 + 2\langle x, y\rangle + \|y\|^2 \leq \|x\|^2 + 2\|x\| \cdot \|y\| + \|y\|^2 = (\|x\| + \|y\|)^2.$ □

Corollary 2.1.10. *Setting* $d(x, y) = \|x - y\|$, (M, d) *is a metric space. Furthermore,* $\|x\| = d(x, 0)$ *is continuous, and thus by 2.1.5,* $\langle x, y\rangle$ *also is continuous in* x *and in* y.

Definition 2.1.11.

(a) *A Banach space is a normed vector space which is complete as a metric space.*
(b) *A Hilbert space is an inner product vector space which is complete as a metric space.*

Remark 2.1.12. Not every norm comes from an inner product. In fact, it can be shown that a norm comes from an inner product iff the parallelogram law 2.1.6 is satisfied (see Appendix B to this chapter). For example, on $\mathbb{R}^2$, defining the norm of a vector $\boldsymbol{v} = (x, y)$ by $\|\boldsymbol{v}\| = |x| + |y|$ will give a norm for which the parallelogram law fails, the "L_1 norm."

Examples of Hilbert spaces 2.1.13.

(a) $\mathbb{R}^n$, $\langle x, y\rangle = \sum_{i=1}^n x_i y_i$.
(b) $\ell_2 = \ell_2(\mathbb{N}) = \{(x_1, x_2, \ldots) : x_i \in \mathbb{R}, \sum_{i=1}^\infty x_i^2 < \infty\}$, and $\langle x, y\rangle = \sum x_i y_i$ which is absolutely convergent since, by the Schwarz inequality,

$$\sum_{i=1}^n |x_i y_i| \leq \left(\sum_{i=1}^n x_i^2\right)^{\frac{1}{2}} \cdot \left(\sum_{i=1}^n y_i^2\right)^{\frac{1}{2}} \leq \left(\sum_{i=1}^\infty x_i^2\right)^{\frac{1}{2}} \cdot \left(\sum_{i=1}^\infty y_i^2\right)^{\frac{1}{2}} = \|x\| \cdot \|y\| \; \forall n,$$

hence

$$\sum_{i=1}^\infty |x_i y_i| \leq \|x\| \cdot \|y\| < \infty.$$

(c) The product of two (or more generally any finite number) Hilbert spaces $M \times N$ with

$$\langle (x_1, y_1), (x_2, y_2)\rangle = \langle x_1, x_2\rangle + \langle y_1, y_2\rangle,$$

or equivalently

$$\|(x, y)\|^2 = \|x\|^2 + \|y\|^2.$$

(d) The collection of holomorphic (complex) functions $f(z)$ on the interior of the unit disc, with $|f(z)|^2$ integrable with respect to planar Lebesgue measure, is an important example of a complex Hilbert space.

Remarks 2.1.14.

(a) $\mathbb{R}^\infty = \{(x_1, x_2, \ldots, 0, 0, \ldots) :$ almost all x_i equal $0\}$ is a dense linear subspace of ℓ_2.
(b) The completeness of ℓ_2 is the Riesz-Fischer theorem, a proof is given in Chapter II Appendix A.
(c) Everything done in this chapter will also work in complex Hilbert spaces, with minor modifications due to the conjugation in $\mathbb{C}$, e.g. the polarization identity is

$$\langle x, y\rangle = \frac{1}{4}(\|x+y\|^2 - \|x-y\|^2 + i\|x+iy\|^2 - i\|x-iy\|^2).$$

2.2 *Orthogonality*

Definition 2.2.1. *A closed linear subspace of a Hilbert space M is called a Hilbert subspace, which we often write (at least in this chapter) H-subspace. In the latter case the H-subspace is obviously itself a Hilbert space with the same norm as M.*

Theorem 2.2.2. *Let Y be an H-subspace of M, $x \in M$, and*

$$\delta = \inf\{\|y - x\| : y \in Y\}.$$

Then there exists a unique $y_0 \in Y$ such that $\delta = d(x, y_0)$, and $(x - y_0) \perp Y$.

Proof. Uniqueness is clear since if y_1 is another such element then

$$(x - y_1) \perp Y \Rightarrow y_1 - y_0 = (x - y_0) - (x - y_1) \perp Y.$$

But $(y_1 - y_0) \in Y$, hence $y_1 - y_0 = 0$. For existence of y_0, let $\{y_n\} \subset Y$ with $\|y_n - x\| \to \delta$. From the Parallelogram law,

$$2\|y_n - x\|^2 + 2\|y_m - x\|^2 = \|y_m + y_n - 2x\|^2 + \|y_n - y_m\|^2.$$

Since $\frac{(y_m+y_n)}{2} \in Y$,

$$\|y_m + y_n - 2x\|^2 = 4\left\|\frac{y_m + y_n}{2} - x\right\|^2 \geq 4\delta^2,$$

thus

$$\|y_n - y_m\|^2 \leq 2\|y_n - x\|^2 + 2\|y_m - x\|^2 - 4\delta^2.$$

As $m, n \to \infty$ the right hand side approaches $2\delta^2 + 2\delta^2 - 4\delta^2 = 0$ which implies that $\{y_n\}$ is Cauchy. Since Y is an H-subspace, $y_n \to y_0 \in Y$. By continuity (2.1.10),

$$\|y_0 - x\| = \lim_{n\to\infty} \|y_n - x\| = \delta.$$

Finally, let $t \in \mathbb{R}$, $y \in Y$, and put $z = y_0 - x$. Then

$$\|z + t\langle y, z\rangle y\| = \|(y_0 + t\langle y, z\rangle y) - x\| \geq \delta = \|z\|$$

implies that

$$0 \leq 2t\langle y, z\rangle^2 + t^2\langle y, z\rangle^2\|y\|^2$$

for all $t \in \mathbb{R}$. Taking t sufficiently small negative implies that $\langle y, z\rangle = 0$. □

Corollary 2.2.3. *Let $X \subset Y \subseteq M$ be a proper inclusion of H-subspaces. Then there exists $y \in Y$, $y \neq 0$, with $y \perp X$. This is true even if Y is just a linear subspace, not necessarily closed in M.*

Definition 2.2.4. *For any subset $A \subset M$, $A^{\perp} = \{x : x \perp A\}$.*

Note that $A^{\perp}$ is always an H-subspace since $\langle\ ,\ \rangle$ is continuous by 2.1.10 and since $A^{\perp}$ is clearly a linear subspace.

Theorem 2.2.5. *Let $X \subseteq M$ be any H-subspace, then*

$$M = X \boxplus X^{\perp} \quad \textit{(internal orthogonal direct sum).}$$

Proof. Suppose $x \in M$. By 2.2.2 there exists a unique $y_0 \in X$ such that $x - y_0 \in X^{\perp}$. Then $x = y_0 + (x - y_0)$ implies that $M = X + X^{\perp}$. Clearly,

$$X \cap X^{\perp} = \{0\},$$

so this is an internal orthogonal direct sum. □

Of course, this is only interesting when $X \subset M$.

Definition 2.2.6. *With the notation of 2.2.5, define the orthogonal projection $\pi_X : M \twoheadrightarrow X$ by $\pi_X(x) = y_0$.*

Proposition 2.2.7.

(a) *π_X is linear,*
(b) *$\pi_X|_X = id_X$,*
(c) *π_X is idempotent, i.e., $\pi_X \circ \pi_X = \pi_X$.*

Proposition 2.2.8. *For any subset $A \subset M$, $A^{\perp} = (\bar{A})^{\perp}$.*

Proof. $A \subseteq \bar{A} \Rightarrow (\bar{A})^{\perp} \subseteq A^{\perp}$, and the reverse implication is immediate from the continuity of $\langle\ ,\ \rangle$. □

2.3 *Bounded linear operators*

Definition 2.3.1. *A linear transformation $f : M \to N$ of Hilbert spaces is called a bounded operator if there exists $\kappa \geq 0$ such that $\|f(x)\| \leq \kappa\|x\|$ for all $x \in M$. In this case, the infimum of all such κ is defined to be $\|f\|$.*

Definition 2.3.2. *For any linear transformation $f : M \to N$,*

$$\text{Ker}\, f = \{x \in M : f(x) = 0\} \subseteq M$$

and

$$\text{Im}\, f = \{y \in N : y = g(x) \text{ for some } x \in M\} \subseteq N.$$

Proposition 2.3.3. *Let $f : M \to N$ be a bounded operator. Then*

(a) *f is continuous,*
(b) *the bounded linear operators and Hilbert spaces form a category $\mathfrak{H}$,*
(c) *Ker f is an H-subspace,*
(d) *Im f is a linear subspace.*

Proof.

(a) $\|f(y) - f(x)\| = \|f(y - x)\| \leq \kappa \cdot \|y - x\|$, where $\kappa = \|f\|$, clearly implies continuity.
(b) $\|id_M\| = 1$, $\|g \circ f\| \leq \|g\| \cdot \|f\|$ suffices to show that id_M is bounded and the composition of two bounded operators is again bounded, i.e. $\mathfrak{H}$ is a category.
(c) and (d): Trivial. □

The fact that Im f is not in general closed (cf. Example 2.3.8(a) below) will be of great importance in the later development of the theory.

Definition 2.3.4. *If $f : M \to \mathbb{R}$ is a bounded operator, then it is called a bounded linear functional (on M). The collection of all such f is written as M^*, the dual space of M.*

Riesz representation theorem 2.3.5. If $f \in M^*$, then there exists a unique $y \in M$ such that $f(x) = \langle x, y\rangle$ for all $x \in M$.

Proof. If Ker $f = M$ then $f = 0$ and $y = 0$ works, so assume that Ker $f \subset M$. Since it is an H-subspace, by Corollary 2.2.3, there exists $y_1 \in (\text{Ker } f)^\perp$, $y_1 \neq 0$, say without loss of generality, $\|y_1\| = 1$. Let $f(y_1) = a \neq 0$ (since $y_1 \notin \text{Ker } f$) and set $y = ay_1$. Then $\|y\| = |a| > 0$ and $f(y) = af(y_1) = a^2 > 0$. For any $x \in M$,

$$f\left(x - \frac{f(x)}{a^2} \cdot y\right) = f(x) - \frac{f(x)}{a^2} \cdot a^2 = 0,$$

hence $x - \frac{f(x)}{a^2} y \in \text{Ker } f \perp y$ and

$$\langle x, y\rangle = \left\langle\left(x - \frac{f(x)}{a^2} y + \frac{f(x)}{a^2} y\right), y\right\rangle = 0 + \frac{f(x)}{a^2}\langle y, y\rangle = f(x). \qquad \square$$

Definition 2.3.6. *Let $\phi : M_1 \to M_2$ be a bounded operator and set $f(x) = \langle \phi x, y\rangle$, where $y \in M_2$. It is easy to see that f is a bounded linear functional, so by the Riesz Representation Theorem 2.3.5, $f(x) = \langle x, \phi^* y\rangle$ for a unique $\phi^* y \in M_1$. Clearly, $\phi^* : M_2 \to M_1$ is linear and ϕ^* is called the adjoint of ϕ.*

Proposition 2.3.7.

(a) *ϕ^* is a bounded operator, with $\|\phi^*\| = \|\phi\|$,*
(b) *$\phi^{**} = \phi$,*
(c) *$(a\phi + b\psi)^* = a\phi^* + b\psi^*$,*
(d) *if ϕ is invertible then $(\phi^{-1})^* = (\phi^*)^{-1}$,*
(e) *Ker $\phi^* = (\text{Im } \phi)^\perp$ (which also is equal to $\overline{\text{Im } \phi}^\perp$ by 2.2.8).*

Examples 2.3.8.

(a) Let $f : \ell_2 \to \ell_2$, $f(x_1, x_2, \ldots) = (\frac{x_1}{1}, \frac{x_2}{2}, \frac{x_3}{3}, \ldots)$, then $\|f\| = 1$, and $\mathbb{R}^\infty \subset \operatorname{Im} f \Rightarrow \operatorname{Im} f$ is dense. But $(1, \frac{1}{2}, \frac{1}{3}, \ldots) \notin \operatorname{Im} f$, so $\operatorname{Im} f$ is not an H-subspace. Here, $\operatorname{Ker} f = \{0\}$.

(b) The orthogonal projection $\pi_X : M \twoheadrightarrow X$ is a bounded operator with $\|\pi_X\| = 1$ (assuming $X \neq \{0\}$), $\operatorname{Im} \pi_X = X$, $\operatorname{Ker} \pi_X = X^\perp$.

(c) Among all linear transformations $f : \ell_2 \to \ell_2$, a cardinality argument can be used to show that "most" are unbounded. But exhibiting an explicity defined (i.e. without the use of transfinite methods) unbounded operator seems difficult.

Lemma 2.3.9. *Let X be a linear subspace of M. Then X is dense iff $X^\perp = \{0\}$ ($= (\bar{X})^\perp$ from 2.2.8).*

Proof. $\Rightarrow$ Let $x \in X^\perp$ with $\|x\| = 1$, supposing $X^\perp \neq \{0\}$. Since X is dense, there exists $y \in X$ with $\|y - x\| < \frac{1}{3}$. But then $x \perp y \Rightarrow \|y - x\|^2 = \|y\|^2 + \|x\|^2 > 1$, a contradiction.

$\Leftarrow$ Since $M = \bar{X} \boxplus (\bar{X})^\perp$ by 2.2.5, and $(\bar{X})^\perp = \{0\}$, we have $M = \bar{X}$, i.e. X is dense. □

The next lemma is not generally included in the basics of Hilbert space, but it will be needed in Chapter III.

Lemma 2.3.10. *If $f : M_1 \rightarrowtail M_2$ is an injective bounded operator, then $\operatorname{Im}(f^* f)$ is dense in M_1.*

Proof. First note that $\operatorname{Im}(f^* f)$ is a linear subspace of M_1. Let $y \in M_1$, $y \perp \operatorname{Im}(f^* f)$. Then $0 = \langle f^* f x, y \rangle = \langle f x, f y \rangle$ for all $x \in M_1$. In particular, $\langle f^* f y, y \rangle = 0 \Rightarrow 0 = \langle f(y), f(y) \rangle = \|f(y)\|^2 \Rightarrow f(y) = 0$ by 2.1.4 $\Rightarrow y = 0$ as f is injective. By Lemma 2.3.9, $\operatorname{Im}(f^* f)$ is dense. □

Definition 2.3.11.

(a) *$f : M \to M$ is self adjoint (symmetric) if $f = f^*$.*

(b) *$f : M \to M$ is orthogonal if f is invertible and $f^{-1} = f^*$. Orthogonality is easily seen to be equivalent to $\|f x\| = \|x\|$ for all $x \in M$, i.e. f is norm preserving (and hence also inner product preserving, i.e. $\langle f x, f y \rangle = \langle x, y \rangle$ for all $x, y \in M$).*

Exercise. If $h : M_1 \to M_2$ is a norm-preserving isomorphism, then one has $h^{-1} = h^*$.

At this stage in a course on Hilbert space, one would soon turn to the study of eigenvalues of a bounded operator f, the collection of eigenvalues being called the spectrum of f (a subset of $\mathbb{C}$), and the spectral theorem for bounded operators that are self adjoint, or more generally that are normal ($f f^* = f^* f$). This would take us too far afield, but we will need the following special case of the spectral theorem. A self adjoint operator f always has real eigenvalues and is called **positive definite** if $\langle f x, x \rangle > 0$ for all $x \neq 0$, or equivalently all eigenvalues are positive real numbers; similarly **positive semi-definite**. The next theorem is essentially the same as that in

[19], p. 265, or can be considered a special case of the "square root lemma for Banach algebras," given as an exercise in [11], p. 243.

Theorem 2.3.12. *Every positive (semi-)definite bounded operator A possesses a positive (semi-)definite square root, denoted* $A^{1/2}$. *It can be represented as the limit (in the strong sense, where* $\{A_n\} \to A$ *in the strong sense means that* $\|A_n x - Ax\| \to 0$ *for all* $x \in M$*) of a sequence of polynomials in A, hence commutes with any bounded transformation that commutes with A, and is unique subject to these conditions.*

Let us give a rough sketch of the proof, the basic idea being due to Isaac Newton! We start with the binomial series

$$(1-t)^{1/2} = 1 - \sum_{n=1}^{\infty} a_n t^n,$$

where $t \in \mathbb{R}, a_1 > a_2 > a_3 > \cdots > 0$. Indeed for $n \geq 2$, one has $a_n = \binom{2n-2}{n} / ((2n-2) \cdot 2^{2n-2})$. It can easily be checked that $\lim_{n\to\infty} a_n = 0$, and furthermore that the series is absolutely convergent, $-1 \leq t \leq 1$.

Now consider A as above. By replacing it with rA for any $r, 0 < r < 1/\|A\|$, we can suppose wlog that $\|A\| < 1$. Set $B = I - A$, then also $\|B\| < 1$. One can now apply the above binomial series, with 1 replaced by I and t replaced by B. The nth partial sum of the expansion is a polynomial of degree n in B (hence also in A), and these polynomials converge in the strong sense to a positive (semi-)definite operator C with $C^2 = A$. This establishes existence of a square root, and indeed only $\|B\| < 1$ was used in this part of the proof, which in turn depends on A being positive (semi)-definite.

The uniqueness proof is simpler, and we will give a very short proof of this for the positive definite case. So suppose also $D^2 = A$, $DA = AD$, and D positive definite. Then D will commute with any polynomial in A, and hence also commute with C which is a strong limit of such polynomials. We therefore have $0 = C^2 - D^2 = C^2 - DC + CD - D^2 = (C - D)(C + D)$. But $C + D$ is also positive definite, so in particular is invertible, whence $C - D = 0$. For the positive semi-definite case cf. [19], p. 264, where a slightly more delicate argument is used.

Remark 2.3.13. We have omitted the discussion of a Hilbert basis for a Hilbert space M, as distinct from an algebraic basis. The idea is clear, e.g. $e_1 = (1, 0, 0, \ldots)$, $e_2 = (0, 1, 0, 0, \ldots)$, ... form a Hilbert basis for ℓ_2 since their linear span is dense in ℓ_2. Thus $\dim_{\mathfrak{H}} \ell_2 = \aleph_0$. In these notes we will only be concerned with separable Hilbert space M, i.e. $\dim_{\mathfrak{H}} M \leq \aleph_0$. A bounded operator is determined by its values on a Hilbert basis, by continuity.

Appendix A

Proof of the Riesz-Fischer theorem

In any metric space, a convergent sequence is Cauchy, so it remains to prove that, conversely, any Cauchy sequence $\{x_n\}$, $x_n = (x_{n1}, x_{n2}, \ldots) \in \ell_2$, converges in ℓ_2. For any fixed i,

$$|x_{ni} - x_{mi}| \leq \|x_n - x_m\|$$

implies that the sequence $x_{1i}, x_{2i}, \ldots$ is a Cauchy sequence in $\mathbb{R}$. So it converges, say to $u_i = \lim_{n\to\infty} x_{ni} \in \mathbb{R}$.

Choose M so that $m, n \geq M \Rightarrow \|x_n - x_m\| \leq \frac{1}{2}$. Then $n \geq M$ implies that

$$\|x_n\| = \|x_M + (x_n - x_M)\| \leq \|x_M\| + \frac{1}{2}. \tag{1}$$

Next write $x_n^{(k)} = (x_{n1}, \ldots, x_{nk}), u^{(k)} = (u_1, \ldots, u_k)$. Since $\lim_{n\to\infty} x_{ni} = u_i$, i.e. $\lim_{n\to\infty} |x_{ni} - u_i| = 0$, one has $\lim_{n\to\infty} \|x_n^{(k)} - u^{(k)}\| = 0$. So there exists N_k with $\|x_n^{(k)} - u^{(k)}\| < \frac{1}{2}, n \geq N_k$. Then $n \geq N_k$ implies that

$$\|u^{(k)}\| = \|x_n^{(k)} + (u^{(k)} - x_n^{(k)})\| < \|x_n^{(k)}\| + \frac{1}{2} \leq \|x_n\| + \frac{1}{2},$$

that is

$$n \geq N_k \Rightarrow \|u^{(k)}\| < \|x_n\| + \frac{1}{2}. \tag{2}$$

Combining (1) and (2), for $n \geq \max\{M, N_k\}$, we have

$$\|u^{(k)}\| < \|x_M\| + 1.$$

Since M does not depend on k, this shows that $\|u^{(k)}\|$ is bounded above, i.e. $u \in \ell_2$.

Finally, choose L so that $m, n \geq L \Rightarrow \|x_n - x_m\| < \frac{\varepsilon}{3}$, for a given $\varepsilon > 0$. Also, it is clearly possible to choose k large enough so that $\|x_L - x_L^{(k)}\| < \frac{\varepsilon}{6}$, $\|u - u^{(k)}\| < \frac{\varepsilon}{6}$. We also have $\|x_n^{(k)} - x_L^{(k)}\| \leq \|x_n - x_L\| < \frac{\varepsilon}{3}$ for all $n \geq L$, whence $\|u^{(k)} - x_L^{(k)}\| \leq \frac{\varepsilon}{3}$ by taking the limit as $n \to \infty$. Combining, we have $n \geq L$ implies that

$$\|u - x_n\| \leq \|u - u^{(k)}\| + \|u^{(k)} - x_L^{(k)}\| + \|x_L^{(k)} - x_L\| + \|x_L - x_n\|,$$

i.e.

$$\|u - x_n\| < \frac{\varepsilon}{6} + \frac{\varepsilon}{3} + \frac{\varepsilon}{6} + \frac{\varepsilon}{3} = \varepsilon.$$

This shows $x_n \to u$. □

Appendix B

Proof of the equivalence of the parallelogram law with norms arising from inner products

As mentioned in 2.1.12, the parallelogram law for a norm is equivalent to the norm coming from an inner product. Although not necessary for our purposes, we shall nevertheless give this clever little proof, due to Frechet, and to Jordan and von Neumann, in the 1930's. One direction (2.1.6) is trivial, so now suppose we have a norm satisfying the parallelogram law. The first step is obvious, define the inner product $\langle x, y\rangle$ using the polarization identity 2.1.5. The properties of being symmetric and strictly positive are immediate, it is only the bilinearity that needs proof. Two other obvious facts are that $\langle -x, z\rangle = -\langle x, z\rangle$, and that $\langle x, x\rangle = \|x\|^2$, i.e. the norm coming from the new inner product is the same as the given norm.

We now show that $\langle x + y, z\rangle = \langle x, z\rangle + \langle y, z\rangle$. We have

$$\begin{aligned}4(\langle x, z\rangle + \langle y, z\rangle) &= \|x+z\|^2 - \|x-z\|^2 + \|y+z\|^2 - \|y-z\|^2\\ &= \|x+z\|^2 + \|y+z\|^2 - (\|x-z\|^2 + \|y-z\|^2).\end{aligned}$$

Let $u = \frac{x+y}{2} + z$, $v = \frac{x-y}{2}$. Then, applying the parallelogram law,

$$\begin{aligned}\|x+z\|^2 + \|y+z\|^2 &= \|u+v\|^2 + \|u-v\|^2 = 2\|u\|^2 + 2\|v\|^2\\ &= 2\left\|\frac{x+y}{2} + z\right\|^2 + 2\left\|\frac{x-y}{2}\right\|^2.\end{aligned}$$

Similarly, $\|x-z\|^2 + \|y-z\|^2 = 2\left\|\frac{x+y}{2} - z\right\|^2 + 2\left\|\frac{x-y}{2}\right\|^2$.

Subtracting these last two equations and applying the first one together with polarization, we obtain

$$4(\langle x, z\rangle + \langle y, z\rangle) = 2\left(\left\|\frac{x+y}{2} + z\right\|^2 - \left\|\frac{x+y}{2} - z\right\|^2\right) = 8\left\langle\frac{x+y}{2}, z\right\rangle.$$

Thus $\langle x, z\rangle + \langle y, z\rangle = 2\langle\frac{x+y}{2}, z\rangle$. Replacing y with 0 and x with $2x$ in this equality gives $\langle 2x, z\rangle = 2\langle x, z\rangle$, so the previous equality can be simplified to $\langle x, z\rangle + \langle y, z\rangle = \langle x + y, z\rangle$, as desired. Also, by an easy induction, this implies $\langle nx, z\rangle = n\langle x, z\rangle$ for all $n \in \mathbb{N}$. It is then easy to argue successively that $\langle rx, z\rangle = r\langle x, z\rangle$, first for $r \in \mathbb{Z}$, then for $r \in \mathbb{Q}$, and finally using continuity (2.1.10) for $r \in \mathbb{R}$. □

3. Hilbert G-modules and von Neumann dimension

In this chapter, most of the algebraic machinery needed for our study of ℓ_2-homology is introduced. The main extra ingredient is a group G that acts freely and isometrically on a Hilbert space. In the topological applications G will turn out to be a quotient group of the fundamental group $\pi_1(X)$, where X is a finite CW-complex. Such a group is always countable, and this will be assumed henceforth, although it is possible to define $\ell_2 G$ even for G uncountable.

3.1 *Definition of $\ell_2 G$ and its module structure*

Definition 3.1.1. *Let G be a countable group, then*

$$\ell_2 G = \left\{ f : G \xrightarrow{f} \mathbb{R},\ \sum_{x\in G} (f(x))^2 < \infty \right\}.$$

For uncountable groups G one would simply use all functions $f : G \to \mathbb{R}$ with "countable support," but for our purposes countable groups suffice. To simplify notation, $\sum$ will denote $\sum_{x \in G}$ unless otherwise noted, and we write $f = \sum f(x)x$. Then $\langle f, g \rangle = \sum f(x)g(x)$, which is absolutely convergent just as in Example 2.1.13(b), and the notation $\ell_2 G$ is consistent with $\ell_2 \mathbb{N}$ as used in that example.

Definition 3.1.2. *The group algebra $\mathbb{R}[G] = \mathbb{R}G \subseteq \ell_2 G$ consists of those functions with finite support, i.e.*

$$\mathbb{R}G = \{r : G \xrightarrow{f} \mathbb{R}, \quad r(x) = 0 \text{ for almost all } x \in G\}.$$

This is clearly a dense linear subspace of $\ell_2 G$ (if G is finite then $\mathbb{R}G = \ell_2 G$) and just like the group ring $\mathbb{Z}G$ it has a multiplication that turns it into the "group algebra,"

$$rs = \left(\sum_{x \in G} r(x)x \right) \cdot \left(\sum_{y \in G} s(y)y \right) = \sum_{x, y \in G} r(x)s(y)xy,$$

all sums being finite (we will generally use letters $r, s \in \mathbb{R}G$). Letting $xy = z$, we can rewrite the above equation as a (left or right) "convolution product"

$$rs = \sum_{z \in G} \left[\sum_{y \in G} r(zy^{-1})s(y) \right] z = \sum_{z \in G} \left[\sum_{x \in G} r(x)s(x^{-1}z) \right] z.$$

Observation 3.1.3. The algebra $\mathbb{R}G$ is associative, has unity $1 = 1 \cdot e$, and is commutative iff G is commutative. Alas, the multiplication in $\mathbb{R}G$ does not in general extend to $\ell_2 G \supseteq \mathbb{R}G$, there are convergence problems as Example 3.1.4 below explicitly shows. However, there is no problem to multiply elements of $\ell_2 G$ (on the left or on the right) by elements of $\mathbb{R}G$ (again, since elements of $\mathbb{R}G$ are *finite* sums). So the correct way to think of $\ell_2 G$ is as an $\mathbb{R}G$-bimodule, and we write $r \cdot f$ $(f \cdot r)$ for the product of an element of $\mathbb{R}G$ with one of $\ell_2 G$ (or vice versa). For $r \in \mathbb{R}G$, we write $L_r : \ell_2 G \to \ell_2 G$ for left multiplication by r, similarly for R_r. The multiplication formula for the (left) module structure is essentially the same as the convolution product formula above, namely for $r \in \mathbb{R}G$ and $f \in \ell_2 G$,

$$r \cdot f = \sum_{z \in G} \left[\sum_{y \in G} r(zy^{-1})f(y) \right] z = \sum_{z \in G} \left[\sum_{x \in G} r(x)f(x^{-1}z) \right] z.$$

Note that the sums on x and on y are finite in this expression, that the right module structure can be similarly defined, and also that $r \cdot s = rs$ when $r, s \in \mathbb{R}G$.

Example 3.1.4. Take $G = \mathbb{Z}$, and define $f, g : G \to \mathbb{R}$ by

$$f(n) = \begin{cases} 0, & n \leq 0 \\ n^{-3/5}, & n \geq 1 \end{cases}, \qquad g(n) = f(-n).$$

Note $\sum_{n\geq 1} n^{-6/5}$ is convergent, so $f, g \in \ell_2 G$. However $f \cdot g \notin \ell_2 G$, indeed

$$(f \cdot g)(n) = \sum_{q=n+1}^{\infty} \frac{1}{q^{3/5}(q-n)^{3/5}} > \int_{n+1}^{\infty} \frac{dx}{[x(x-n)]^{3/5}}$$

$$> \int_{n+1}^{\infty} \frac{dx}{x^{6/5}} = \frac{5}{(n+1)^{1/5}},$$

showing that $f \cdot g$ is not square summable.

Proposition 3.1.5.

(a) *For $r, s \in \mathbb{R}G$, $r \cdot s = rs$ (left or right action),*
(b) *G acts by isometries, i.e. for $r = 1 \cdot y$, both L_y and R_y are isometries of $\ell_2 G$,*
(c) *For $r, s \in \mathbb{R}G$ and $f \in \ell_2 G$, $r \cdot (s \cdot f) = (r \cdot s) \cdot f$, $(f \cdot s) \cdot r = f \cdot (s \cdot r)$. Furthermore, $r \cdot (f \cdot s) = (r \cdot f) \cdot s$ and $\ell_2 G$ is thus an $\mathbb{R}G$-bimodule.*

Proof.

(a) This holds by definition.
(b) Note that for $r = 1 \cdot y = y$, $y \in G$, one has

$$r \cdot f = y \cdot f = \sum_{z\in G} 1 \cdot f(y^{-1}z)z = \sum_{z\in G} f(y^{-1}z)z.$$

The result is now obvious since $\{f(y^{-1}x) : x \in G\} = \{f(x) : x \in G\}$.
(c) $y \cdot \left(z \cdot \sum f(x)x\right) = y \cdot \sum f(z^{-1}x)x = y \cdot \sum g(x)x$, where $g(x) := f(z^{-1}x)$. Thus, we get

$$y \cdot \left(z \cdot \sum f(x)x\right) = \sum g(y^{-1}x)x = \sum f(z^{-1}y^{-1}x)x = \sum f((yz)^{-1}x)x,$$

i.e. $y \cdot (z \cdot \sum f(x)x) = (yz) \cdot \sum f(x)x$ holds for $y, z \in G$, and extends by linearity to all $r, s \in \mathbb{R}G$. The proof for the bimodule structure is similar, one shows that $(y \cdot f) \cdot z = \sum f(y^{-1}xz^{-1})x = y \cdot (f \cdot z)$ and extends linearly. □

Proposition 3.1.6. *The action of $\mathbb{R}G$ on $\ell_2 G$ (left or right) is by bounded operators.*

Proof. Let $r = \sum r(y)y \in \mathbb{R}G$, $f \in \ell_2 G$. Using 3.1.5(b), we have $\|y \cdot f\| = \|f\|$, thus

$$\|r \cdot f\| = \left\|\sum r(y)y \cdot f\right\| \leq \sum \|r(y)y \cdot f\| = \sum |r(y)|\,\|y \cdot f\| = \sum |r(y)|\,\|f\|$$

which implies that $\|L_r\| \leq |r| := \sum |r(y)| < \infty$, being a finite sum. □

Notice that $(\ell_2 G)^n$ also becomes an $\mathbb{R}G$-bimodule with bounded operators and G acting by isometries, where $r \cdot (f_1, \ldots, f_n) = (r \cdot f_1, \ldots, r \cdot f_n)$, the so-called diagonal action. One frequently says G-module or left G-module, omitting the $\mathbb{R}$, similarly for G-invariant and G-equivariant. We close this section with four useful technical results.

Proposition 3.1.7. *Let M be any Hilbert space that is a (left) $\mathbb{R}G$-module with G acting by isometries (e.g. $M = (\ell_2 G)^n$). If V is a G-invariant Hilbert subspace, then so is $V^{\perp}$.*

Proof. Since $V^\perp$ is a Hilbert subspace, it suffices to show $v \in V^\perp \Rightarrow y \cdot v \in V^\perp$, for $y \in G$. But for any $w \in V$,

$$\langle w, y \cdot v\rangle = \langle y \cdot y^{-1} \cdot w, y \cdot v\rangle = \langle y^{-1} \cdot w, v\rangle = 0$$

since $y^{-1} \cdot w \in V$. □

Corollary 3.1.8. *If V is a G-invariant Hilbert subspace, then π_V is G-equivariant.*

Proof. Let $f = v + w \in M$ where $v \in V, w \in V^\perp$ (uniquely). For any $y \in G$, $y \cdot f = y \cdot v + y \cdot w$ with $y \cdot v \in V$ and (by 3.1.7) $y \cdot w \in V^\perp$. Then by definition

$$\pi_V(y \cdot f) = y \cdot v = y \cdot \pi_V(f).$$ □

Note that $\mathbb{R}G \otimes_{\mathbb{R}G} \ell_2 G \approx \ell_2 G$, so that $(\mathbb{R}G)^n \otimes_{\mathbb{R}G} \ell_2 G \approx (\ell_2 G)^n$ since tensor product distributes over direct sums. We then have

Proposition 3.1.9. *Let $\phi : (\mathbb{R}G)^n \to (\mathbb{R}G)^m$ be a morphism of (right) $\mathbb{R}G$-modules. Then the induced operator $\tilde{\phi}$ defined as the composition*

$$(\ell_2 G)^n \approx (\mathbb{R}G)^n \otimes_{\mathbb{R}G} \ell_2 G \xrightarrow{\phi\otimes\mathrm{id}} (\mathbb{R}G)^m \otimes_{\mathbb{R}G} \ell_2 G \approx (\ell_2 G)^m$$

is bounded.

Proof. The $m \times n$ matrix $[\phi_{ij}]$ of ϕ satisfies

$$\phi(a_1, \dots, a_n) = \left(\sum_{j=1}^{n} \phi_{1j} a_j, \dots, \sum_{j=1}^{n} \phi_{mj} a_j\right) \qquad a_j, \phi_{ij} \in \mathbb{R}G.$$

Write (as usual) $\phi_{ij} = \sum \phi_{ij}(x)x$ (finite sum) and $|\phi_{ij}| := \sum |\phi_{ij}(x)|$. Then for $f \in \ell_2 G$, one has $\|\phi_{ij} \cdot f\| \le |\phi_{ij}|\,\|f\|$ just as in 3.1.6, so

$$\|\tilde{\phi}(f_1, \dots, f_n)\|^2 = \sum_i \left\|\sum_j (\phi_{ij} \cdot f_j)\right\|^2 \le \sum_{i,j} |\phi_{ij}|^2 \|f_j\|^2,$$

i.e.

$$\|\tilde{\phi}(f_1, \dots, f_n)\|^2 \le \sum_{i,j} |\phi_{ij}|^2 \|(f_1, \dots, f_n)\|^2 = \left(\sum_{i,j} |\phi_{ij}|^2\right) \|(f_1, \dots, f_n)\|^2.$$

□

Similarly, given a morphism ψ of left $\mathbb{R}G$-modules, one uses $\mathrm{id} \otimes_{\mathbb{R}G} \psi$ to define the bounded operator $\tilde{\psi}$.

Proposition 3.1.10. *If $f : M_1 \to M_2$ is a G-equivariant bounded operator of Hilbert $\mathbb{R}G$-modules on which G acts by isometries, then*

(a) *so is $f^* : M_2 \to M_1$,*
(b) *so is $g : M_1 \to M_1$, where $g^2 = f^* f$, g is self adjoint and positive definite.*

Proof.

(a) Let $\alpha \in G$, $y \in M_2$. For any $x \in M_1$, $\langle x, f^*(\alpha \cdot y)\rangle = \langle f(x), \alpha \cdot y\rangle = \langle \alpha^{-1} \cdot f(x), (\alpha^{-1}\alpha) \cdot y\rangle = \langle f(\alpha^{-1} \cdot x), y\rangle = \langle \alpha^{-1} \cdot x, f^*(y)\rangle = \langle x, \alpha \cdot f^*(y)\rangle$. Thus $\langle x, f^*(\alpha \cdot y) - \alpha \cdot f^*(y)\rangle = 0$, so by 2.1.4, $f^*(\alpha \cdot y) = \alpha \cdot f^*(y)$.
(b) $\langle x, f^* f x\rangle = \langle fx, fx\rangle \geq 0$ implies that $f^* f$ is self adjoint and positive semi-definite. So, g exists by Theorem 2.3.12. From (a), $g^2 = f^* f$ is G-equivariant and again, by Theorem 2.3.12, the same is true for g. □

3.2 *Hilbert G-modules*

We start by formalizing some of the properties, proved in §3.1 for $(\ell_2 G)^n$, into the following definition.

Definition 3.2.1.

(a) *A Hilbert G-module M is a left $\mathbb{R}G$-module M which is a Hilbert space on which G acts by isometries such that M is isometrically G-isomorphic to a G-invariant subspace of $(\ell_2 G)^n$, for some n.*
(b) *Morphisms of Hilbert G-modules are the bounded G-equivariant operators $f : M_1 \to M_2$, and this forms a category $\mathfrak{H}G$.*

Quotient modules 3.2.2. Let M be a Hilbert G-module and $V \subset M$ a G-invariant linear subspace. Then $\bar{V}$ is G-invariant and $M/\bar{V}$ has a natural Hilbert space structure with

$$\|\,[w]\,\| = \inf\{\|w - v\| : v \in \bar{V}\} = \|w - \pi_{\bar{V}}(w)\|, \qquad w \in M.$$

Furthermore, $\pi_{\bar{V}}$ induces, by restriction to $V^\perp$, a G-equivariant isometric isomorphism of Hilbert G-modules $V^\perp \overset{\approx}{\to} M/\bar{V}$.

Definition 3.2.3. *A map $f : M_1 \to M_2$ of Hilbert G-modules is a*

(a) *weak isomorphism if f is injective, bounded, G-equivariant, Im f is dense in M_2,*
(b) *strong isomorphism if f is an isometric G-equivariant isomorphism of Hilbert spaces.*

The next theorem is a little surprising, and quite useful.

Theorem 3.2.4. *If $f : M_1 \to M_2$ is a weak isomorphism, then there exists a strong isomorphism $h : M_1 \to M_2$.*

Proof. As in the proof of 3.1.10(b), $f^* f$ is self adjoint positive semi-definite, indeed positive definite since f is injective. It also has $\operatorname{Im}(f^* f)$ dense by Lemma 2.3.10. So, as in 3.1.10, there exists a positive definite self adjoint operator g with $g^2 = f^* f$, and (also by 3.1.10) g is G-equivariant. Furthermore, $\operatorname{Im} g \supseteq \operatorname{Im} g^2 = \operatorname{Im}(f^* f)$ and hence is dense. Since g is injective, $g^{-1} : \operatorname{Im} g \to M_1$, which is bijective, exists; we set $\tilde{h} = f \circ g^{-1} : \operatorname{Im} g \to M_2$. Then $\operatorname{Im} \tilde{h} = \operatorname{Im} f$ is dense in M_2, and using $g^* = g$, for any $x, y \in \operatorname{Im} g$, we have

$$\langle \tilde{h}x, \tilde{h}y\rangle = \langle fg^{-1}(x), fg^{-1}(y)\rangle = \langle f^* fg^{-1}(x), g^{-1}(y)\rangle = \langle g^2 g^{-1}(x), g^{-1}(y)\rangle,$$

i.e.

$$\langle \tilde{h}x, \tilde{h}y\rangle = \langle gg^{-1}(x), g^*g^{-1}(y)\rangle = \langle x, gg^{-1}(y)\rangle = \langle x, y\rangle.$$

Hence $\tilde{h} : \operatorname{Im} g \to \operatorname{Im} f$ is an isometric isomorphism, and since $\operatorname{Im} g \subseteq M_1$, $\operatorname{Im} f \subseteq M_2$ are dense, $\tilde{h}$ extends by continuity to an isometric isomorphism $h : M_1 \to M_2$. Since f and g are already known to be G-equivariant, so are (successively) g^{-1}, $\tilde{h}$, h, and thus h is a strong isomorphism of Hilbert G-modules. □

Definition 3.2.5. *Two Hilbert G-modules M_1 and M_2 are isomorphic ($M_1 \approx M_2$) if there exists a weak isomorphism $M_1 \to M_2$.*

By Theorem 3.2.4, the existence of a weak isomorphism implies the existence of a strong isomorphism, thus this is an equivalence relation. As a second application, we have the following:

Proposition 3.2.6. *Let $\phi : M_1 \to M_2$ be a bounded G-equivariant operator of Hilbert G-modules. Then* $(\operatorname{Ker}\phi)^\perp \approx M_1/\operatorname{Ker}\phi \approx \overline{\operatorname{Im}\phi}$.

Proof. The first isomorphism follows from 3.2.2. The composition $i \circ \rho$ where $i : \operatorname{Im}\phi \hookrightarrow \overline{\operatorname{Im}\phi}$ is the inclusion map, and $\rho : M_1/\operatorname{Ker}\phi \to \operatorname{Im}\phi$ is the standard bijective map, is a weak isomorphism, so the second isomorphism follows from Theorem 3.2.4. □

3.3 *von Neumann dimension*

3.3.1. Our goal in this section is to define an "equivariant" dimension, called the von Neumann dimension, $\dim_G M$, of a Hilbert G-module M satisfying

(a) $\dim_G M \in \mathbb{R}_0^+$,
(b) $\dim_G M = 0$ iff $M = 0$,
(c) $\dim_G M = \dim_G N$ if $M \approx N$,
(d) $\dim_G(M \oplus N) = \dim_G M + \dim_G N$,
(e) $M \subseteq N \Rightarrow \dim_G M \leq \dim_G N$,
(f) $\dim_G(\ell_2 G) = 1$,
(g) G finite $\Rightarrow \dim_G M = \frac{1}{|G|}\dim_{\mathbb{R}} M$,
(h) If H is a subgroup of G with finite index then $\dim_G M = \dim_H M/[G : H]$.

Remarks 3.3.2.

(a) This idea goes back to the 1936 paper of Murray and von Neumann [17], and is closely related to what they call the centre-valued trace.
(b) For G finite, both (f) and (h) follow from (g).
(c) From (d) and (f), $\dim_G(\ell_2 G)^n = n$.
(d) For $G = \{e\}$, $\ell_2 G = \mathbb{R}$ and $\dim_G M = \dim_{\mathbb{R}} M$, i.e. the theory reduces to ordinary linear algebra.

Definition 3.3.3. *The Kaplansky trace map $\rho : \mathbb{R}G \to \mathbb{R}$ is given by $\rho(\sum r(x)x) = r(e)$.*

This seemingly simple idea will turn into a powerful tool, as we shall soon see. Some obvious properties are $\rho(e) = 1, \rho(r+s) = \rho(r) + \rho(s)$, and $\rho(rs) = \sum_y r(y^{-1})s(y) = \sum_x r(x)s(x^{-1}) = \rho(sr)$. Note that ρ can equally well be defined on $\ell_2(G)$, with the same properties except for the last (since $\ell_2(G)$ has no product).

Definition 3.3.4. *The von Neumann algebra $N(G) = \hom_{\mathfrak{H}G}(\ell_2 G, \ell_2 G)$ is the algebra of bounded left G-equivariant operators $\ell_2 G \to \ell_2 G$.*

Definition 3.3.5. *Conjugation on $\ell_2 G$ (or $\mathbb{R}G$ by restriction) is the map $f = \sum f(x)x \longmapsto \bar{f} = \sum f(x)x^{-1}$.*

Notice that $\rho(f) = \rho(\bar{f})$. Next recall $\ell_2 G$ is an $\mathbb{R}G$-bimodule. The right action of $\mathbb{R}G$ on $\ell_2 G$ is by bounded left G-equivariant operators (check that $L_y R_z = R_z L_y \in N(G)$, this means that R_z is left G-equivariant), so in this sense $\mathbb{R}G \subseteq N(G)$ as a subalgebra (specifically $z \mapsto R_z$, $z \in \mathbb{R}G$, note then that $e \mapsto I = \mathrm{id}_{NG}$). The adjoint map $\phi \longmapsto \phi^*$ gives an involution on $N(G)$ which turns it into a real C^*-algebra. In $\ell_2 G$ the adjoint of R_y is easily seen to be $R_{y^{-1}}$, as shown by

$$\langle u, R_y(v)\rangle = \langle u, vy\rangle = \langle uy^{-1}, vyy^{-1}\rangle = \langle uy^{-1}, v\rangle \Rightarrow R_y^* = R_{y^{-1}}.$$

This shows that under the inclusion $\mathbb{R}G \subseteq NG$, conjugation in $\mathbb{R}G$, as defined in 3.3.5, corresponds to the adjoint in NG.

We now extend the Kaplansky trace map to a trace on NG, as follows.

Definition 3.3.6. *Let $\phi \in NG$, then $\mathrm{tr}_G(\phi) := \langle \phi(e), e\rangle \in \mathbb{R}$.*

Proposition 3.3.7.

(a) *If $\phi \in \mathbb{R}G$ then $\mathrm{tr}_G(\phi) = \rho(\phi)$,*
(b) $\mathrm{tr}_G(\phi) = \mathrm{tr}_G(\phi^*)$, *where $\phi \in NG$,*
(c) $\mathrm{tr}_G(I) = 1$, $\mathrm{tr}_G(\phi + \psi) = \mathrm{tr}_G(\phi) + \mathrm{tr}_G(\psi)$, $\phi, \psi \in NG$,
(d) $\mathrm{tr}_G(\phi\psi) = \mathrm{tr}_G(\psi\phi)$, $\phi, \psi \in NG$,
(e) *Given a non-zero G-invariant Hilbert subspace $V \subseteq \ell_2 G$, then* $0 < \mathrm{tr}_G(\pi_V) \leq 1$,
(f) *If one replaces $\phi \in \mathbb{R}G$ by $\phi \in \mathbb{Q}G$ (resp. $\mathbb{Z}G$) in the above definitions, then* $\mathrm{tr}_G(\phi) \in \mathbb{Q}$ *(resp.* $\mathrm{tr}_G(\phi) \in \mathbb{Z}$).

Proof.

(a) Set $\phi = \sum r(x)x$, then $\mathrm{tr}_G(\phi) = \langle e \cdot \sum r(x)x, e\rangle = \langle \sum r(x)x, e\rangle = r(e) \cdot 1 = r(e)$.
(b) $\mathrm{tr}_G(\phi) = \langle \phi(e), e\rangle = \langle e, \phi^*(e)\rangle = \mathrm{tr}_G(\phi^*)$.
(c) Trivial.
(e) The hypotheses on V imply $\pi_V \in NG$. Now $e = \pi_V(e) + y$ with $\pi_V(e) \perp y$. Then

$$\mathrm{tr}_G(\pi_V) = \langle \pi_V(e), e\rangle = \langle \pi_V(e), \pi_V(e)\rangle = \|\pi_V(e)\|^2.$$

This implies, by the Pythagorean theorem, that $0 \leq \mathrm{tr}_G(\pi_V) \leq 1$. To see that it cannot equal 0, let $0 \neq f \in V$. Then $f(x) \neq 0$ for some $x \in G$, and now define $g := x^{-1} \cdot f$, we have $g \in V$ since V is G-invariant. Then $g(e) = (x^{-1} \cdot f)(e) = f(x) \neq 0$ shows that V is not orthogonal to e, whence $\pi_V(e) \neq 0$.
(f) Clear for the Kaplansky trace ρ, then use (a).
(d) Given in Chapter III, Appendix A. □

The definition of the von Neumann dimension will now be briefly indicated; establishing then that it is well defined and all the properties given in 3.3.1 is not difficult but will be omitted here to remain within the time constraints of the mini-course (however, see Chapter III, Appendix B).

Definition 3.3.8.

(a) *Let $M_n(N(G))$ be the algebra of bounded left G-equivariant operators $(\ell_2 G)^n \to (\ell_2 G)^n$ (thus, $M_1(N(G)) = N(G)$). Any operator $F \in M_n(N(G))$ gives rise in the usual way to an $n \times n$ matrix $[F_{ij}]$, where each $F_{ij} \in N(G)$. Then*

$$\mathrm{tr}_G(F) := \sum_{i=1}^{n} \mathrm{tr}_G(F_{ii}).$$

(b) *Let $V \subseteq (\ell_2 G)^n$ be a G-invariant Hilbert subspace. By Corollary 3.1.8, $\pi_V \in M_n(N(G))$. Then*

$$\dim_G V := \mathrm{tr}_G(\pi_V) \in \mathbb{R}.$$

Definition 3.3.9. *Let M be an arbitrary Hilbert G-module and choose a G-equivariant isometric isomorphism $\alpha : M \overset{\approx}{\to} V \subseteq (\ell_2 G)^n$. Then*

$$\dim_G(M) := \dim_G V.$$

Remark 3.3.10. It is a classical result that $\mathbb{Z}[G]$ has no idempotents apart from 0, 1 (for a proof see Chapter III, Appendix A). However, in $\mathbb{R}[G]$, where say $G = C_2 = \{1, t\}$, $r = \frac{1}{2}(1 + t)$ is idempotent.

Kaplansky Conjecture. If G is torsion free then $\mathbb{R}[G]$ has no non-trivial idempotents.

In 1972 Zalesskii [21] showed $\mathrm{tr}_G(\epsilon) \in \mathbb{Q}$ for ϵ idempotent in $\mathbb{R}[G]$ and G torsion free. Strengthening this from $\mathbb{Q}$ to $\mathbb{Z}$ would prove the conjecture. Further work in this direction was done in 1998 by Burger and Valette [2].

Appendix A

In this appendix, among other things, the proofs of 3.3.7(d) and Remark 3.3.10, are given.

Proof of 3.3.7(d). Let $\phi \in NG$. For all $x \in G$, we have

$$\langle \phi^*(e), x \rangle = \langle e, \phi(x) \rangle = \langle e, x\phi(e) \rangle = \langle x^{-1}, \phi(e) \rangle = \langle \bar{x}, \phi(e) \rangle,$$

using the fact that G acts by isometries for the second last equality. Further, using the evident fact that $\langle f, g \rangle = \langle \bar{f}, \bar{g} \rangle$ for $f, g \in \ell_2 G$, we have

$$\langle \bar{x}, \phi(e) \rangle = \langle \phi(e), \bar{x} \rangle = \overline{\langle \phi(e), x \rangle}.$$

It follows that $\phi^*(e) = \overline{\phi(e)}$. Using this, as well as the conjugation identity just mentioned, we obtain

$$\begin{aligned}\mathrm{tr}_G(\phi\psi) &= \langle \phi(\psi(e)), e\rangle = \langle \psi(e), \phi^*(e)\rangle \\ &= \langle \psi(e), \overline{\phi(e)}\rangle = \langle \overline{\psi(e)}, \phi(e)\rangle \\ &= \langle \psi^*(e), \phi(e)\rangle = \langle e, \psi(\phi(e))\rangle \\ &= \mathrm{tr}_G(\psi\phi).\end{aligned}$$

□

A first application of 3.3.7(d) appears in the following lemma, which is then used to prove Remark 3.3.10.

Lemma. *Suppose* $\epsilon \in NG$ *with* $\epsilon^2 = \epsilon$. *Let* $V = \overline{\mathrm{Im}\,\epsilon}$. *Then* $\mathrm{tr}_G(\pi_V) = \mathrm{tr}_G(\epsilon)$, *and* $\mathrm{tr}_G(\epsilon) + \mathrm{tr}_G(I - \epsilon) = 1$.

Proof. The second result is a trivial consequence of $\epsilon + (I - \epsilon) = I$, together with 3.3.7(c). To establish the first, let us write $\pi = \pi_V$ for convenience. Since $\epsilon^2 = \epsilon$, $\epsilon|(\mathrm{Im}\,\epsilon) = \mathrm{id}_{\mathrm{Im}\,\epsilon}$. Continuity then implies $\epsilon|V = \mathrm{id}_V$, whence $\epsilon \circ \pi = \pi$. Similarly, since $\pi|V = \mathrm{id}_V$, we have $\pi \circ \epsilon = \epsilon$. Applying 3.3.7(d), $\mathrm{tr}_G(\epsilon) = \mathrm{tr}_G(\pi \circ \epsilon) = \mathrm{tr}_G(\epsilon \circ \pi) = \mathrm{tr}_G(\pi)$. □

Corollary. $0 \leq \mathrm{tr}_G(\epsilon) \leq 1$, *and* $0 \leq \mathrm{tr}_G(I - \epsilon) \leq 1$.

Proof. Simply apply 3.3.7(e) to the results of the lemma. □

Proposition. *The only idempotents in* $\mathbb{Z}G$ *are* 0 *and* e.

Proof. Let ϵ be an idempotent in $\mathbb{Z}G \subset \mathbb{R}G \subseteq NG$. By the corollary, $0 \leq \mathrm{tr}_G(\epsilon) \leq 1$. But by 3.3.7(f) this number is now an integer, so either $\mathrm{tr}_G(\epsilon) = 0$ or $\mathrm{tr}_G(\epsilon) = 1$. In the first case, by 3.3.7(e) (using also $\mathrm{tr}_G(\epsilon) = \mathrm{tr}_G(\pi_V)$), we have $V = \{0\}$ and hence $\epsilon = 0$. In the second case, the same reasoning shows $I - \epsilon = 0$, giving $\epsilon = I$. □

Appendix B

In this appendix we show that the definition of von Neumann dimension, as given in 3.3.9, is well defined. Once that is done the other properties in 3.3.1 are relatively straightforward to establish, we omit these verifications. Appendix A will be used in the proof.

Proposition. *The von Neumann dimension* $\dim_G(M)$ *of a Hilbert* G*-module* M, *as given by Definition 3.3.9, is well defined.*

Proof. Recall that a G-equivariant isometric isomorphism $\alpha : M \overset{\approx}{\rightarrow} V \subseteq (\ell_2 G)^m$ was first chosen, so suppose $\beta : M \overset{\approx}{\rightarrow} W \subseteq (\ell_2 G)^n$ is another such. We may suppose wlog that $n \geq m$, say $n = m + k, k \geq 0$. Letting V' denote V considered as a subspace of $(\ell_2 G)^n = (\ell_2 G)^m \oplus (\ell_2 G)^k$ under the inclusion $z \mapsto (z, 0)$, one easily sees from Definition 3.3.8(b) that $\dim_G(V') = \dim_G(V)$. Thus, replacing α by α

followed by this inclusion, we may assume wlog that $m = n$ and $V = V'$. Now define the G-equivariant isometric isomorphism $h := \beta \circ \alpha^{-1} : V \to W$, and extend h to an operator $H \in M_n(NG)$ by putting $H|V = h$, $H|V^{\perp} = 0$. By the Exercise after 2.3.11, $h^*h = I_V$. It follows that $H^*H = \pi_V$, and similarly $HH^* = \pi_W$. Then, using Proposition 3.3.7(d) (proved in Appendix A), we have

$$\dim_G(V) = \mathrm{tr}_G(H^*H) = \mathrm{tr}_G(HH^*) = \dim_G(W).$$

It follows that $\dim_G(M)$ is well defined. □

4. Real homology of finite complexes and harmonic chains

In this chapter the ordinary homology of finite CW-complexes with real coefficients, $H_*(X;\mathbb{R})$, is considered from a slightly novel point of view, using harmonic chains. This approach will make the introduction of ℓ_2-homology, in Chapter V, relatively straightforward. As a "preview," consider the usual short exact sequence defining the ith homology groups of X,

$$0 \to B_i \hookrightarrow Z_i \to H_i \to 0.$$

With real coefficients this becomes

$$0 \to B_i(X;\mathbb{R}) \hookrightarrow Z_i(X;\mathbb{R}) \to H_i(X;\mathbb{R}) \to 0,$$

a short exact sequence of vector spaces which necessarily splits, i.e. (supressing the $\mathbb{R}$ in the notation) $Z_i \approx B_i \oplus H_i$ as an external direct sum. Then also $Z_i = B_i \boxplus (B_i)^{\perp}$ as an internal orthogonal direct sum, with $B_i^{\perp} \approx H_i$. We shall study $B_i^{\perp} := \mathcal{H}_i$, the so-called harmonic chains.

4.1 *Harmonic chains*

Let X be a finite CW-complex with (integral) cellular chain complex $(K_*(X), d)$. Its real chain complex is then $C_* := \mathbb{R} \otimes K_*(X)$, with $1 \otimes d$ as differential, also written d. Let $\sigma_1, \ldots, \sigma_{\alpha_i}$ be the i-cells of X, then these form the natural basis for $C_i(X)$ (which we often write simply C_i). Thus

Proposition 4.1.1. *$C_i(X)$ is a finite dimensional real Hilbert space of dimension α_i, with orthonormal basis $\{\sigma_1, \ldots, \sigma_{\alpha_i}\}$ and associated inner product $\langle\, ,\, \rangle : C_i \times C_i \to \mathbb{R}$.*

Definition 4.1.2. $\delta_{i-1} = d_i^* : C_{i-1}(X) \to C_i(X)$.

Thus, $\langle \delta_{i-1}x, y\rangle = \langle x, d_i y\rangle$ (also $\langle \delta_i x, y\rangle = \langle x, d_{i+1}y\rangle$). This is equivalent to the next result (or use Proposition 2.3.7(e)).

Proposition 4.1.3. *Ker* $d_i = (\mathrm{Im}\,\delta_{i-1})^{\perp}$, *and Ker* $\delta_i = (\mathrm{Im}\, d_{i+1})^{\perp}$.

Definition 4.1.4. $Z_i = \operatorname{Ker} d_i$, $B_i = \operatorname{Im} d_{i+1}$, $Z^i = \operatorname{Ker} \delta_i$, $B^i = \operatorname{Im} \delta_{i-1}$, *all subspaces of* C_i. *We call* Z_i, B_i, Z^i, B^i *respectively the* i*-cycles,* i*-boundaries,* i*-cocycles,* i*-coboundaries.*

Proposition 4.1.5.

(a) $B_i \subseteq Z_i, \quad B^i \subseteq Z^i$,
(b) $C_i = B^i \boxplus Z_i = B_i \boxplus Z^i$,
(c) $B^i \perp B_i$.

Proof.

(a) It follows from $d_i d_{i+1} = 0$ and hence also $0 = (d_{i-1}d_i)^* = d_i^* d_{i-1}^* = \delta_i \delta_{i-1}$.
(b) From Theorem 2.2.5, $C_i = \operatorname{Im} \delta_{i-1} \boxplus (\operatorname{Im} \delta_{i-1})^\perp$, and using 4.1.3, 4.1.4 $C_i = B^i \boxplus Z_i$. Similarly for $B_i \boxplus Z^i$.
(c) It follows from $\langle \delta_{i-1}x, d_{i+1}y\rangle = \langle x, d_i d_{i+1} y\rangle = 0$ for all $x, y \in C_i$. □

Corollary 4.1.6. $C_i = B_i \boxplus B^i \boxplus (Z_i \cap Z^i)$.

Proof. $C_i = B_i \boxplus Z^i = B_i \boxplus (Z^i \cap C_i) = B_i \boxplus (Z^i \cap (B^i \boxplus Z_i)) = B_i \boxplus B^i \boxplus (Z_i \cap Z^i)$, using $B^i \subseteq Z^i$ in the final step. □

Definition 4.1.7. *The harmonic* i*-chains of* X *are*

$$\mathcal{H}_i(X) := Z_i(X) \cap Z^i(X).$$

Thus, $C_i = B_i \boxplus B^i \boxplus \mathcal{H}_i$, called the Hodge-de Rham decomposition of $C_i(X)$. The following diagram is a useful mnemonic for this orthogonal decomposition of $C_i(X)$.

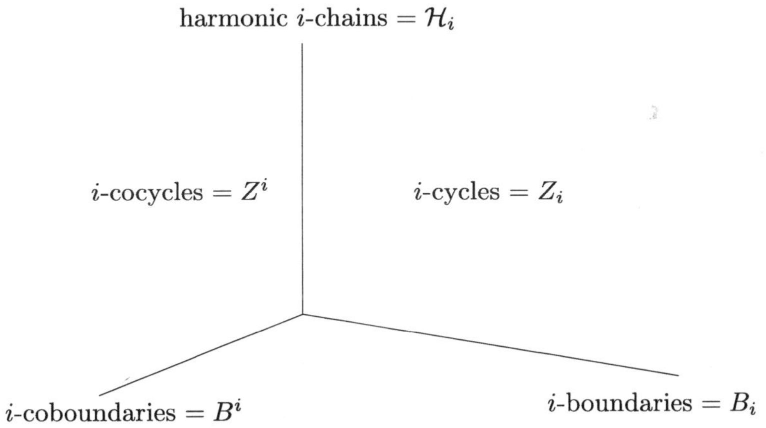

Definition 4.1.8. *The Laplacian is* $\Delta_i := d_{i+1}\delta_i + \delta_{i-1}d_i : C_i \to C_i$.

Proposition 4.1.9. $\mathcal{H}_i = \operatorname{Ker} \Delta_i$.

Proof. $\mathcal{H}_i \subseteq \operatorname{Ker} \Delta_i$ is clear. Conversely, suppose $\Delta_i x = 0$, then $d_{i+1}\delta_i(x) = -\delta_{i-1}d_i(x)$ in $B_i \cap B^i = \{0\}$. This implies that $d_{i+1}\delta_i(x) = \delta_{i-1}d_i(x) = 0 \Rightarrow \delta_i(x) \in B^{i+1} \cap Z_{i+1} = \{0\}$ and $d_i(x) \in B_{i-1} \cap Z^{i-1} = \{0\}$. So $x \in \operatorname{Ker} \delta_i \cap \operatorname{Ker} d_i = \mathcal{H}_i$. □

Exercise. Show Δ_i is self-adjoint, positive semi-definite, and $\Delta_i^2 = 0$.

4.2 *Euler characteristic and Morse inequalities*

Definition 4.2.1.

$$\chi(X) = \sum_{i \geq 0} (-1)^i \alpha_i$$

is the (usual) Euler characteristic of X.

Definition 4.2.2. $b_i(X) = \dim_{\mathbb{R}} \mathcal{H}_i(X)$ *is the (not quite usual) ith Betti number of* X.

In fact, the usual definition of Betti number is $\dim_{\mathbb{R}} Z_i - \dim_{\mathbb{R}} B_i$.

Proposition 4.2.3. $b_i(X) = \dim_{\mathbb{R}} Z_i - \dim_{\mathbb{R}} B_i$.

Proof. Similar to the proof of 4.1.6, we have $Z_i = Z_i \cap C_i = Z_i \cap (B_i \boxplus Z^i) = B_i \boxplus \mathcal{H}_i$ implies that $\dim_{\mathbb{R}} Z_i = \dim_{\mathbb{R}} B_i + \dim_{\mathbb{R}} \mathcal{H}_i$. □

Thus $b_i(X)$ equals in fact the usual Betti number, implying the following theorem found in every algebraic topology text.

Theorem 4.2.4. $\chi(X) = \sum_{i \geq 0} (-1)^i b_i(X)$.

Similarly we have, with essentially the same proof

Morse inequalities 4.2.5. For any $k \geq 0$,

$$\alpha_k - \alpha_{k-1} + \alpha_{k-2} - \cdots + (-1)^k \alpha_0 \geq b_k - b_{k-1} + \cdots + (-1)^k b_0.$$

Proof. Consider $C_{i-1} = B_{i-1} \boxplus Z^{i-1} \overset{\delta_i}{\twoheadrightarrow} B^i$. Since $\operatorname{Ker} \delta_i = Z^{i-1}$, $\delta_i : B_{i-1} \overset{\approx}{\rightarrow} B^i$, so $\dim_{\mathbb{R}} B^i = \dim_{\mathbb{R}} B_{i-1}$. Now $C_i = B_i \boxplus B^i \boxplus \mathcal{H}_i$ implies that $\alpha_i = \dim B_i + \dim B_{i-1} + b_i$, i.e. $(\alpha_i - b_i) = \dim B_i + \dim B_{i-1}$. Then

$$\sum_{i=0}^{k} (-1)^{k-i} (\alpha_i - b_i) = (\dim B_k + \dim B_{k-1}) - (\dim B_{k-1} + \dim B_{k-2}) + \cdots$$

or $\sum_{i=0}^{k} (-1)^{k-i} (\alpha_i - b_i) = \dim B_k \geq 0$. □

Notice that by taking $k \geq \dim K$, this also proves 4.2.4, since B_k is then 0.

4.3 *Homology and cohomology*

4.3.1. We have already noted in the proof of 4.2.3 that $Z_i = B_i \boxplus \mathcal{H}_i$. Since $H_i(X; \mathbb{R}) = Z_i / B_i$, we have $H_i(X; \mathbb{R}) \approx \mathcal{H}_i$ with the isomorphism induced by $\pi_{H_i} : Z_i \twoheadrightarrow \mathcal{H}_i$.

4.3.2. For cohomology one uses the cochain complex

$$C^* = C^*(X) = \hom_{\mathbb{R}}(C_*(X), \mathbb{R}),$$

with differential $\delta^{i-1} = \hom_{\mathbb{R}}(d_i, 1) : C^{i-1} \to C^i$. The inner product of C_* induces a natural isomorphism

$$\Lambda_i : C_i \to C^i X$$

where $\sigma \longmapsto \langle\sigma, \rangle$ for each i-cell σ of X. Since $(\Lambda_{i+1})(\sigma)(c) = \langle\delta_i\sigma, c\rangle = \langle\sigma, d_{i+1}c\rangle = (\Lambda_i\sigma)(d_{i+1}c) = (\delta^i\Lambda_i)(\sigma)(c)$, $\Lambda_* : (C_*, \delta_*) \to (C^*, \delta^*)$ defines an isomorphism of cochain complexes with $Z^i(X) \overset{\approx}{\to} \operatorname{Ker}\delta^i$, $B^i(X) \overset{\approx}{\to} \operatorname{Im}\delta^{i-1}$. This justifies the terminology Z^i = cocycles, B^i = coboundaries, that we have been using, even though Z^i, $B^i \subseteq C_i$.

Definition 4.3.3. *For $f : X \to Y$, using the isomorphism in 4.3.1, we define $f_! : \mathcal{H}_i X \to \mathcal{H}_i Y$ as the composition*

$$\mathcal{H}_i X \overset{\approx}{\leftarrow} H_i(X;\mathbb{R}) \overset{f_*}{\to} H_i(Y;\mathbb{R}) \overset{\approx}{\to} \mathcal{H}_i Y.$$

This makes $\mathcal{H}_i$ a covariant functor on the category of finite CW-complexes and continuous maps, also $f_!$ depends only on the homotopy class of f.

Similarly one can define $f^! : \mathcal{H}_i Y \to \mathcal{H}_i X$ using

$$\mathcal{H}_i Y \overset{\approx}{\to} H^i(Y;\mathbb{R}) \overset{f^*}{\to} H^i(X;\mathbb{R}) \overset{\approx}{\leftarrow} \mathcal{H}_i X.$$

This makes $\mathcal{H}_i$ into a contravariant functor, one easily checks that $f_!$, $f^!$ are adjoints.

5. Infinite complexes and ℓ_2-homology

The preliminary work in the previous chapters will now reap its dividends. With minor modifications of the definitions in Chapter IV, the ℓ_2-chains and ℓ_2-homology of an infinite (or finite) CW-complex Y will now be defined, taking into account a group action of some group G on Y, where the action is free, cocompact, and cellular, i.e. G permutes the cells of Y. The ℓ_2-chains (homology) will all be Hilbert G-modules.

5.1 *Description of the ℓ_2-chains*

Regular cellular coverings 5.1.1. The general situation considered from now on is that of a group G acting freely and cellularly on a connected CW-complex Y. In this case, denoting the orbit CW-complex by $X = Y/G$, the covering projection $p : Y \twoheadrightarrow X$ is a regular covering, i.e. $p_*\pi_1(Y, y_0)$ is a normal subgroup of $\pi_1(X, x_0)$. We also assume the action of G on Y to be cocompact, so that X is a compact (hence finite) CW-complex.

Example 5.1.2. A simple but useful example to keep in mind is the standard universal covering projection $p : \mathbb{R} \twoheadrightarrow S^1$, with $G = \mathbb{Z}$ acting on $\mathbb{R}$ by translations by integers. Similarly, the universal cover of any compact CW-complex can serve as an example.

Remarks 5.1.3.

(a) Since X is finite, $\pi_1(X)$ is countable, hence so is $G = \pi_1(X)/p_*\pi_1(Y)$ (cf. the introduction to Chapter III).
(b) Also note that in this situation the ordinary (integral) chain group $K_i(Y)$ is a finitely generated free module over the group ring $\mathbb{Z}G$, with rank α_i equal to the number of i-cells of X, and d_i is a $\mathbb{Z}G$ map.

Definition 5.1.4. $C_i^{(2)}(Y) := \{\sum_{\sigma\in J_i} f(\sigma)\sigma : f(\sigma) \in \mathbb{R}, \sum_{\sigma\in J_i}(f(\sigma))^2 < \infty\}$, *the square summable chains of* Y, *where* J_i *is the set of* i*-cells of* Y. *Clearly* $C_i^{(2)}(Y)$ *is a Hilbert space with orthonormal basis* J_i.

Definition 5.1.5. *The* ℓ_2*-chains of* Y *are*

$$C_i(Y, G) = \ell_2 G \otimes_{\mathbb{Z}G} K_i(Y),$$

which we often write simply $C_i(Y)$ *if no confusion is possible. Note that since* $\ell_2 G$ *is a (left)* $\mathbb{R}G$*-module, so is* $C_i(Y)$ *via*

$$r \cdot (f \otimes_{\mathbb{Z}G} x) := (r \cdot f) \otimes_{\mathbb{Z}G} x, \quad r \in \mathbb{R}G, \ f \in \ell_2 G, \ x \in K_i(Y).$$

The group G permutes the set J_i of i-cells, choose from each orbit a representative $\bar{\tau}^i_\mu$, $\mu \in 1, \ldots, \alpha_i$, where α_i is the number of i-cells of X. The collection $\{x \otimes \bar{\tau}^i_\mu : x \in G, \mu \in \{1, \ldots, \alpha_i\}\}$ then may be taken as an orthonormal basis for $C_i(Y)$. The Hilbert space structure thus induced on $C_i(Y)$ is independent of the choice of representatives $\bar{\tau}^i_\mu$. Indeed, we have

Proposition 5.1.6. $C_i(Y)$ *and* $C_i^{(2)}(Y)$ *are naturally isomorphic as Hilbert spaces.*

Proof. The canonical bijection $\{x \otimes \bar{\tau}^i_\mu\} \to J_i$, given by $x \otimes \bar{\tau}^i_\mu \mapsto x \cdot \bar{\tau}^i_\mu$, defines a canonical bijection between the respective orthonormal Hilbert bases of $C_i(Y)$ and $C_i^{(2)}(Y)$, and hence extends to a natural isomorphism of Hilbert spaces. □

Note also that if $f \in \ell_2 G$ then $\|f \otimes \bar{\tau}^i_\mu\| = \left\| \sum f(x)x \otimes \bar{\tau}^i_\mu \right\| = \|f\|$, proving

Proposition 5.1.7. *The map* $(\ell_2 G)^{\alpha_i} \to C_i Y$, $(f_1, \ldots, f_{\alpha_i}) \mapsto \sum_{\mu=1}^{\alpha_i} f_\mu \otimes \bar{\tau}^i_\mu$, *defines an isometric* G*-equivariant isomorphism of Hilbert spaces.*

Definition 5.1.8. *The boundary map on* $C_*(Y)$ *is*

$$C_i(Y) = \ell_2 G \otimes_{\mathbb{Z}G} K_i(Y) \stackrel{\ell_2 G \otimes d_i}{\longrightarrow} \ell_2 G \otimes_{\mathbb{Z}G} K_{i-1}(Y) = C_{i-1}(Y).$$

Taking the $\mathbb{Z}G$ *bases* $\{\bar{\tau}^i_\mu : \mu = 1, \ldots, \alpha_i\}$ *and* $\{\bar{\tau}^{i-1}_\nu : \nu = 1, \ldots, \alpha_{i-1}\}$ *for* $K_i(Y)$ *and* $K_{i-1}(Y)$ *respectively,* d_i *(being* G*-equivariant) represents a morphism* $(\mathbb{Z}G)^{\alpha_i} \to (\mathbb{Z}G)^{\alpha_{i-1}}$ *of* $\mathbb{Z}G$*-modules, so also via the inclusion* $\mathbb{Z} \subseteq \mathbb{R}$ *a morphism* $(\mathbb{R}G)^{\alpha_i} \to (\mathbb{R}G)^{\alpha_{i-1}}$ *of* $\mathbb{R}G$*-modules.*

Proposition 5.1.9. *The boundary map* $\ell_2 G \otimes_{\mathbb{Z}G} d_i$ *is a bounded operator.*

Proof. It is identical to the map

$$\ell_2 G \otimes_{\mathbb{R}G} d_i : \ell_2 G \otimes_{\mathbb{R}G} K_i(Y; \mathbb{R}) \to \ell_2 G \otimes_{\mathbb{R}G} K_{i-1}(Y; \mathbb{R});$$

now apply 3.1.9. □

We usually write simply d_i for $\ell_2 G \otimes_{\mathbb{R}G} d_i$.

Example 5.1.10. We return to Example 5.1.2, so let $\mathbb{Z} \approx G = \{z^n : n \in \mathbb{Z}\}$. Then, following Lück, Chapter 1, $\mathbb{R}G = \mathbb{R}[z, z^{-1}]$, the Laurent polynomials over $\mathbb{R}$, $\ell_2 G \approx L^2(S^1)$, the square integrable functions on S^1 with respect to Lebesgue measure μ on S^1, $NG \approx L^\infty(S^1)$, the essentially bounded measurable functions $S^1 \to \mathbb{R}$, and $\mathrm{tr}_G(f) = \int f d\mu$.

5.2 *Unreduced and reduced ℓ_2-homology*

We have seen in §5.1 that the operators $d_i : C_i(Y) \to C_{i-1}(Y)$ are bounded (hence continuous) and G-equivariant. The same then holds for their adjoints $\delta_{i-1} = d_i^*$ (cf. 2.3.7(a) and 3.1.10(a)). As in §4.1 put $\operatorname{Ker} d_i = Z_i(Y) = Z_i$, $\operatorname{Ker} \delta_i = A^i$, and $\mathcal{H}_i = Z_i \cap Z^i$; these are all G-invariant Hilbert subspaces of $C_i = C_i(Y)$. Similarly define $B_i = \operatorname{Im} d_{i+1}$, $B^i = \operatorname{Im} \delta_{i-1}$; these are G-invariant linear subspaces of C_i but in general are not closed. Then, just as in §4.1, we have the next two results (see also the diagram below).

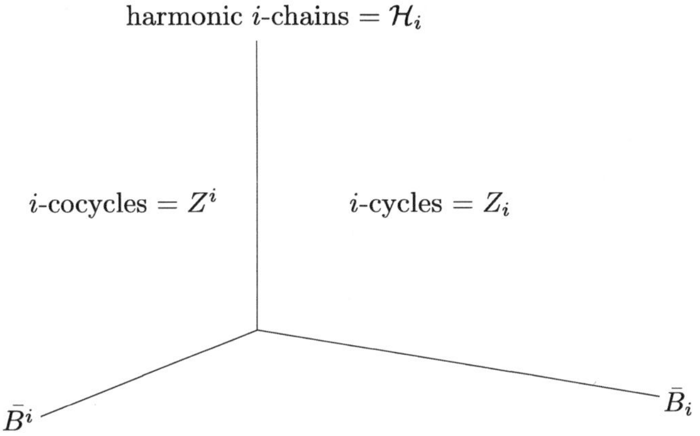

ℓ_2-Hodge-de Rham decomposition 5.2.1. $C_i = \bar{B}^i \boxplus Z_i = \bar{B}_i \boxplus Z^i = \bar{B}^i \boxplus \bar{B}_i \boxplus \mathcal{H}_i$.

Proposition 5.2.2. *$\mathcal{H}_i$ =Ker Δ_i, where $\Delta_i = d_{i+1}\delta_i + \delta_{i-1}d_i$ is the ℓ_2-Laplacian (note $\mathcal{H}_i = \mathcal{H}_i(Y, G)$ is sometimes written for extra clarity).*

The proofs are identical to those in §4.1, apart from a little extra care, using continuity, to first establish that the ℓ_2-analogue of Proposition 4.1.5, with B_i, B^i replaced respectively by $\bar{B}_i$, $\bar{B}^i$, is valid.

Definition 5.2.3.

(a) $H_i(Y) = Z_i / B_i$,
(b) ${}^{red}H_i(Y) = Z_i / \bar{B}_i$.

Caution 5.2.4. The reduced ℓ_2-homology groups ${}^{red}H_i(Y)$ have nothing in common with the notion of reduced homology $\tilde{H}_i(Y)$ in usual homology theory.

Proposition 5.2.5. *${}^{red}H_i(Y) \approx \mathcal{H}_i(Y, G)$, induced by $\pi_{\mathcal{H}_i} : Z_i \to \mathcal{H}_i$.*

Proof. Same as 4.3.1, with B_i replaced by $\bar{B}_i$. □

5.2.6. Similarly, define ${}^{red}H^i(Y) = Z^i / \bar{B}^i$, and again one has ${}^{red}H_i(Y) \approx \mathcal{H}_i(Y, G)$.

Remark 5.2.7. For G finite $B_i = \bar{B}_i$, $B^i = \bar{B}^i$ and everything reduces to the situation of 4.3, with $\mathcal{H}_i(Y, G) \approx H_i(Y; \mathbb{R}) \approx H^i(Y; \mathbb{R})$.

Remark 5.2.8. C_i, Z_i, Z^i, $\bar{B}_i$, $\bar{B}^i$, $\mathcal{H}_i$, and ${}^{red}H_i(Y)$, ${}^{red}H^i(Y)$ are all clearly Hilbert G-modules, C_i being isomorphic to $(\ell_2 G)^{\alpha_i}$ and the others being G-invariant submodules of C_i or quotients of G-invariant submodules.

The non-reduced (co-)homology groups are generally less useful and more difficult to compute. They are not in general Hilbert G-modules.

Definition 5.2.9.

(a) *For the non-reduced ℓ_2-homology,*

$$Z_i/B_i \approx H_i(\ell_2 G \otimes_{\mathbb{Z}G} K_*(Y)) := H_i^G(Y; \ell_2 G),$$

the equivariant homology of Y with coefficients in the G-module $\ell_2 G$.

(b) $H^i_G(Y; \ell_2 G) := H^i(C^*(Y)) = H^i(\hom_{\mathbb{Z}G}(K_*(Y), \ell_2 G)$ *is the equivariant cohomology of Y with coefficients in $\ell_2 G$.*

For equivariant homology, the inclusion $B_i \hookrightarrow \bar{B}_i$ induces a natural surjection $H_i^G(Y; \ell_2 G) \twoheadrightarrow {}^{red}H_i(Y)$.

Remark 5.2.10. For some applications, it is useful to note that everything done in this section generalizes to the case of a regular covering $Y \twoheadrightarrow X$ where the k-skeleton $X^{(k)}$ of X is finite, provided we consider $C_i(Y)$ only for $i < k$.

Definition 5.2.11. *The "canonical" map $can_i\colon H_i(Y; \mathbb{R}) \to {}^{red}H_i(Y)$ is defined by simply considering any ordinary real cycle as an ℓ_2-cycle (with finite support).*

Definition 5.2.12. *The "canonical" map $can^i\colon {}^{red}H^i(Y) \to H^i(Y; \mathbb{R})$ is defined as follows. Consider $x = [\xi] \in {}^{red}H^i(Y) = Z^i/\bar{B}^i$, since $Z^i = \bar{B}^i \boxplus \mathcal{H}_i$, $\xi = \gamma + \eta$ uniquely with $\gamma \in \bar{B}^i$, $\eta \in \mathcal{H}_i$. Then $\xi - \eta = \gamma \in \bar{B}^i \Rightarrow x = [\xi] = [\eta]$, i.e. x has a unique harmonic cocycle representative $\eta \in \mathcal{H}_i$. As in 4.3.2, η identifies under the isomorphism Λ_* with an ordinary cocycle in $C^*(Y; \mathbb{R})$, which thus defines a cohomology class $can^i(x) \in H^i(Y; \mathbb{R})$.*

6. Properties of ℓ_2-homology

6.1 *G-homotopy invariance*

Probably the most important property of $\mathcal{H}_i(Y)$ is that up to isomorphism (as a Hilbert G-module), it depends only on the G-homotopy type of Y. By 5.2.5, this is equivalent to showing the same for ${}^{red}H_i(Y)$. After giving the proof, at the start of this chapter, we look at other properties of ℓ_2-homology including some computations, then study ℓ_2-chain complexes, and finally define and study the ℓ_2-Betti numbers.

Lemma 6.1.1. *${}^{red}H_i$ is a functor from the category of free cocompact G-CW complexes and G-homotopy classes of maps, to the category $\mathfrak{H}G$.*

Proof. Let $f : Y \to Z$ be a G-map of free cocompact G-CW-complexes. By G-cellular approximation $f \simeq_G g$, a cellular G-map. By 3.1.9, $g_* = g_i : C_i(Y) \to$

$C_i(Z)$ are bounded operators, and the g_i also as usual are chain maps. By continuity $g_i(\bar{B}_i(Y)) \subseteq \bar{B}_i(Z)$ so it induces ${}^{red}H_i(g) : {}^{red}H_i(Y) \to {}^{red}H_i(Z)$. If $h : Y \to Z$ is a cellular G-map G-homotopic to g, then $K_i g, K_i h : K_i Y \to K_i Z$ are chain homotopic morphisms of G-chain complexes. Hence $\ell_2 G \otimes_{\mathbb{R}G} K_* g := g_*, \ell_2 G \otimes_{\mathbb{R}G} K_* h := h_*$ are chain homotopic as well. Thus $(g_* - h_*)(Z_i(Y)) \subseteq B_i(Y) \subseteq \bar{B}_i(Y)$, whence ${}^{red}H_i(g_*) = {}^{red}H_i(h_*)$ for all i, showing that ${}^{red}H_i(g_*)$ depends only on the G-homotopy class of f. □

Corollary 6.1.2. *The Hilbert G-modules $\mathcal{H}_i(Y)$ are also functorial and give rise to G-homotopy invariants.*

Corollary 6.1.3. *If $f : Y \to Z$ is a G-map between free cocompact G-CW-complexes and is also a homotopy equivalence, then $\mathcal{H}_i(Y) \approx \mathcal{H}_i(Z)$.*

Proof. Indeed f induces a weak equivalence ${}^{red}H_i(Y) \to {}^{red}H_i(Z)$, thus ${}^{red}H_i(Y) \approx {}^{red}H_i(Z)$ and $\mathcal{H}_i(Y) \approx \mathcal{H}_i(Z)$, all as Hilbert G-modules. □

Remark. There is no need, in this last corollary, to assume that f is a G-homotopy equivalence; it is well known that any G-map between free G-CW-complexes which is a homotopy equivalence is also a G-homotopy equivalence.

The situation for ${}^{red}H^i(Y) = Z^i(Y)/\bar{B}^i(Y)$ is similar, giving contravariant functors with ${}^{red}H^i(f)$ being induced by the adjoint $f_i^* : C_i(Z) \to C_i(Y)$. One also has ${}^{red}H_i(Y) \approx \mathcal{H}_i(Y) \approx {}^{red}H^i(Y)$ as well.

The next simple lemma will be useful in the examples to follow.

Lemma 6.1.4. *If G is infinite then for $n \geq 1$ the left G-module $(\ell_2 G)^n$ contains no G-invariant element besides* 0.

Proof. If $f = \sum_{x \in G} f(x)x \in \ell_2 G$ is G-invariant, then for each $y \in G$ $f = y \cdot f = \sum f(y^{-1}x)x \Rightarrow f(x) = f(y^{-1}x)$ for all $y \in G$ which implies that $f(x)$ is constant. Since $\sum (f(x))^2$ converges and G is infinite, $f(x) = 0$ for all $x \in G$, i.e. $f = 0$. Similarly for $n > 1$. □

Example 6.1.5. Suppose Y is a connected G-CW complex with cocompact 1-skeleton and $|G| = \infty$. Then ${}^{red}H_0(Y) = \mathcal{H}_0(Y) = {}^{red}H^0(Y) = 0$. To see this, first note that $K_1(Y) \overset{d_1}{\to} K_0(Y) \overset{\epsilon}{\twoheadrightarrow} \mathbb{Z} \to 0$ is exact, and since $\otimes$ is a right exact functor

$$C_1(Y) \overset{d_1}{\to} C_0(Y) \twoheadrightarrow \ell_2 G \otimes_G \mathbb{Z} \to 0$$

is also exact. Hence $(\operatorname{Im} d_1)^\perp = \operatorname{Ker} \delta_0 \subseteq C_0(Y)$ is mapped injectively (indeed isomorphically) to $\ell_2 G \otimes_G \mathbb{Z}$, whence $\operatorname{Ker} \delta_0$ consists of G-invariant elements and is 0 by Lemma 6.1.4.

For any discrete group G, its classifying space BG is an Eilenberg-MacLane space $K(G, 1)$, and its universal cover is written EG. We shall usually refer to such a $K(G, 1)$ space as *aspherical*. The projection $EG \to K(G, 1)$ is an example of a regular covering with free cellular action of G on EG, but in general $K(G, 1)$ need not be compact. However, if G is finitely generated, then we can (and do) take a cellular decomposition for $K(G,1)$ with a single 0-cell, and a single 1-cell for each generator of G, thus the 1-skeleton is a finite wedge of circles. Similarly, if G is finitely presented, we take $K(G,1)$ with 1-skeleton as above and with a single 2-cell for each relation, giving a finite $K(G,1)^{(2)}$. We return to these ideas in §7.1. The Whitehead Conjecture is also related to these ideas. It asserts that any subcomplex of an aspherical 2-dimensional complex is also aspherical, and only special cases of it have thus far been proved.

For any finitely generated infinite group G, Example 6.1.5 applies and $\mathcal{H}_0(EG) = 0$. On the other hand, if G is finite, the proof of 6.1.4 shows that the G-invariant elements of $\ell_2 G$ are the constants, i.e. $\mathbb{R}$, and thus $\mathcal{H}_0(EG) \approx \mathbb{R}$.

In the next example we see that in general, in the same situation, the unreduced $H_0(Y)$ is not 0.

Example 6.1.6. Consider now the same example in the Introduction, $Y = \mathbb{R} \twoheadrightarrow X = S^1$, $G = \mathbb{Z}$, $S^1 = e^0 \cup e^1$. In this case the ℓ_2-chain complex $C_*(Y)$ is

$$0 \to \ell_2(\mathbb{Z}) \xrightarrow{d_1} \ell_2(\mathbb{Z}) \to 0.$$

Let x generate $\mathbb{Z} = \{x^n\}$, write $f \in \ell_2(\mathbb{Z})$ as $f = \sum_{n\in\mathbb{Z}} a_n x^n$, $a_n \in \mathbb{R}$. Since $d_1(\bar{e}^1_n) = \bar{e}^0_{n+1} - \bar{e}^0_n = (x-1)\bar{e}^0_n$, $d_1(f) = (x-1)f$. Then clearly d_1 is injective (same argument as in 6.1.4), and from Example 6.1.5 ${}^{red}H_0(Y) = 0$ implies that $\operatorname{Im} d_1$ is dense in $\ell_2(\mathbb{Z})$. But d_1 is not surjective, e.g. $1 \notin \operatorname{Im} d_1$. To see this, if $1 \in \operatorname{Im} d_1$ then $1 = (x-1)\sum a_i x^i$ implies that all a_j, $j < 0$, are equal and hence 0, whence $a_{-1} - a_0 = 1$ implies that $a_0 = -1 = a_1 = a_2 = \cdots$, which is impossible. Therefore $H_0^G(Y; \ell_2 G) \neq 0$, in contrast to ${}^{red}H_0(Y) = 0$.

Equivalently, this example shows that $\mathcal{H}_i(E\mathbb{Z}) = 0$, $i \geq 0$, whereas $H_i^{\mathbb{Z}}(E\mathbb{Z}; \ell_2\mathbb{Z}) = H_i(\mathbb{Z}; \ell_2\mathbb{Z}) = 0$ for $i > 0$ but $H_0(\mathbb{Z}; \ell_2\mathbb{Z}) \neq 0$. For cohomology, $H^1(\mathbb{Z}; \ell_2\mathbb{Z}) \neq 0$ and otherwise $H^i(\mathbb{Z}; \ell_2\mathbb{Z}) = 0$, $i \neq 1$. For a more systematic study of the difference between reduced and unreduced ℓ_2-homology, the reader is referred to the initial work of Novikov and Shubin [18] and its further developments by Farber [7] and Lück [15]. See also §7.3 for relations of this example to amenability.

6.2 $\ell_2 G$-chain complexes

Definition 6.2.1. *A chain complex*

$$V_* : \cdots \to V_{i+1} \xrightarrow{d_{i+1}} V_i \xrightarrow{d_i} V_{i-1} \to \cdots$$

of Hilbert G-modules V_i is called an ℓ_2-chain complex if each d_i is a bounded G-equivariant operator. Of course $d_i d_{i+1} = 0$ for all i also holds (the chain complex condition).

Definitions 6.2.2. *With V_* as above,*

(a) ${}^{red}H_i(V_*) = \operatorname{Ker} d_i / \overline{\operatorname{Im} d_{i+1}}$,

(b) *V_* is called weak exact if ${}^{red}H_i(V_*) = 0\, \forall\, i$.*

Definitions 6.2.3. *Let V_*, W_* be ℓ_2-chain complexes.*

(a) *A morphism $\phi_* : V_* \to W_*$ is an ordinary chain map with each ϕ_i a bounded G-equivariant operator,*
(b) *two morphisms ϕ_*, ψ_* are $\ell_2 G$-homotopic if they are chain homotopic by a chain homotopy of of bounded G-equivariant operators,*
(c) *two $\ell_2 G$-chain complexes are $\ell_2 G$-chain equivalent if there exist $\phi_* : V_* \to W_*$ and $\psi_* : W_* \to V_*$, both morphisms as in (a), with $\psi_*\phi_* \simeq id_{V_*}$, $\phi_*\psi_* \simeq id_{W_*}$ in the sense of (b).*

Just as in the ordinary homology theory, a morphism $\phi_* : V_* \to W_*$ induces well defined bounded G-equivariant operators ${}^{red}H_i(V_*) \to {}^{red}H_i(W_*)$ depending only on the $\ell_2 G$-homotopy class of ϕ_*. The proof is identical to the usual one apart from the extra use of continuity to show that $\phi_i(\overline{\operatorname{Im} d_{i+1}}) \subseteq \overline{\operatorname{Im} d'_{i+1}}$, where d, d' are the respective boundary operators.

Corollary 6.2.4. *If V_*, W_* are ℓ_2-chain equivalent ℓ_2-chain complexes, then ${}^{red}H_i(V_*) \approx {}^{red}H_i(W_*)$ as Hilbert G-modules, for all i.*

Definition 6.2.5. *A weak exact $\ell_2 G$-chain complex $0 \to U \stackrel{\alpha}{\to} V \stackrel{\beta}{\to} W \to 0$ is called a short weak-exact sequence (of Hilbert G-modules). Then α is injective, $\overline{\operatorname{Im}\beta} = W$, and* $\operatorname{Ker}\beta = \overline{\operatorname{Im}\alpha}$.

Proposition 6.2.6. *For a short weak-exact sequence of Hilbert G-modules $0 \to U \stackrel{\alpha}{\to} V \stackrel{\beta}{\to} W \to 0$, $\dim_G V = \dim_G U + \dim_G W$.*

Proof. From 3.2.6, $\overline{\beta(V)} \approx (\operatorname{Ker}\beta)^\perp \approx V/\operatorname{Ker}\beta$, hence

$$\dim_G V = \dim_G(\operatorname{Ker}\beta) + \dim_G(\operatorname{Ker}\beta)^\perp = \dim_G(\overline{\operatorname{Im}\alpha}) + \dim_G(\overline{\operatorname{Im}\beta}),$$

i.e. $\dim_G(V) = \dim_G U + \dim_G W$, since α is injective and $\overline{\operatorname{Im}\beta} = W$. □

Corollary 6.2.7. *Let $V_* : 0 \to V_n \stackrel{d_n}{\to} V_{n-1} \stackrel{d_{n-1}}{\to} \dots V_0 \stackrel{d_0}{\to} 0$ be a chain complex of Hilbert G-modules. Then*

$$\sum_{i\geq 0}(-1)^i \dim_G V_i = \sum_{i\geq 0} \dim_G {}^{red}H_i(V_*).$$

Proof. Set $K_i = \operatorname{Ker} d_i$, $I_i = \overline{\operatorname{Im} d_{i+1}}$. Then we have short weak-exact sequences of Hilbert G-modules

$$0 \to K_i \hookrightarrow V_i \to I_{i-1} \to 0,$$

$$0 \to I_i \hookrightarrow K_i \to {}^{red}H_i(V_*) \to 0.$$

Hence, $\dim_G V_i = \dim_G K_i + \dim_G I_{i-1}$, and $\dim_G {}^{red}H_i = \dim_G K_i - \dim_G I_i$, and the result follows easily (as for ordinary chain complexes). □

6.3 ℓ_2-Betti numbers

Definition 6.3.1. *Let Y be a free cocompact G-CW complex. The ith ℓ_2-Betti number of Y (with respect to G) is*

$$\beta_i(Y; G) := \dim_G {}^{red}H_i(Y).$$

Applying results from §3.3 and §6.1, 6.2 gives immediately

Proposition 6.3.2.

(a) *$\beta_i(Y, G)$ is a G-homotopy invariant of Y, and therefore a homotopy invariant of $X = Y/G$,*
(b) *If H is a subgroup of G with $[G : H] = m < \infty$, then $\beta_i(Y; H) = m\beta_i(Y; G)$ (here $Y/H \twoheadrightarrow Y/G$ is an m-sheeted covering projection),*
(c) *If $|G| < \infty$, then $\beta_i(Y; G) = \frac{1}{|G|} b_i(Y)$, in particular if Y is connected then $\beta_0(Y; G) = \frac{1}{|G|}$,*
(d) *If $|G| = \infty$ and Y is connected then $\beta_0(Y; G) = 0$ (cf. 6.1.5).*

Definition 6.3.3. *Let X be a connected finite CW-complex. Then we define the ℓ_2-Betti number of X by $\beta_i(X) = \beta_i(\tilde{X}; G)$, where $\tilde{X}$ is the universal covering space of X and $G = \pi_1(X, x_0)$.*

Here are two very simple properties.

Proposition 6.3.4. $0 \leq \beta_i(X) \leq \alpha_i$.

Proof. Clear since ${}^{red}H_i(\tilde{X})$ is a subquotient of $(\ell_2 G)^{\alpha_i}$, as Hilbert G-modules. □

Proposition 6.3.5. *If $\hat{X}$ is a connected m-sheeted covering of X with $H = \pi_1(\hat{X})$, $G = \pi_1(X)$, then $\beta_i(\hat{X}) = m\beta_i(X)$.*

Proof. $\beta_i(\hat{X}) = \dim_H {}^{red}H_i(\hat{X}) = m \cdot \dim_G {}^{red}H_i(\hat{X}) = m \cdot \beta_i(X)$. □

Example 6.3.6. Note that this is very different from the behaviour of ordinary Betti numbers, e.g. $S^1 \xrightarrow{s} S^1$, $s(z) = z^2$, is a 2-sheeted covering but $b_1(S^1) = 1 = b_1(\hat{X})$ and $b_1(X) = b_1(S^1) = 1$. From 6.1.6, $\beta_i(S^1) = 0$, for all $i \geq 0$.

Example 6.3.7. More generally, if X (like S^1) is a connected finite CW-complex which possesses a regular finite covering space $\hat{X} \to X$ of degree $m > 1$, with $\hat{X}$ homotopy equivalent to X, then $\beta_i(X) = 0$ for all i, since $\beta_i(X) = \beta_i(\hat{X}) = m \cdot \beta_i(X)$.

Atiyah's conjecture 6.3.8. In [1] Atiyah asked whether the ℓ_2-Betti numbers are rational, or even integral in the case of a torsion free group G.

A somewhat weaker conjecture is the

Zero-divisor conjecture. For G torsion free, $\mathbb{Q}[G]$ contains no non-trivial zero divisors.

It is not very difficult to show the first conjecture implies the second. The zero-divisor conjecture has been proved for a large class of torsion free groups, cf. [14].

We close this section with theorems on the Euler characteristic and Morse inequalities involving ℓ_2-Betti numbers. The proofs are analogous to those in §4.2, using the machinery in §6.2, and are omitted.

Theorem 6.3.9. *For X a finite connected CW-complex,*

$$\chi(X) = \sum_{i \geq 0} (-1)^i \beta_i(X).$$

Theorem 6.3.10. *For X as in 6.3.9, let N be a normal subgroup of $\pi_1(X)$ with $\pi_1(X)/N = Q$. Let $\hat{X}$ denote the covering space associated with N. Then*

$$\chi(X) = \sum_{i \geq 0} (-1)^i \beta_i(\hat{X}; Q).$$

Morse inequalities 6.3.11. Let X be a connected CW-complex with finite $(k+1)$-skeleton. Then

$$\alpha_k - \alpha_{k-1} + \cdots + (-1)^k \alpha_0 \geq \beta_k - \beta_{k-1} + \cdots + (-1)^k \beta_0.$$

7. Applications

At the end of Chapter VI, applications have already been made (of the ℓ_2-Betti numbers) to Euler characteristic and Morse inequalities. Further applications and a few simple computations will be sketched in the present chapter. Proofs and many details will often be omitted here, but it is hoped that the reader will at least get a taste of some of the significant applications that are possible.

7.1 *Two examples*

Example 7.1.1. This example extends Example 6.1.6, $X = S^1$, to the wedge of k circles $X = \bigvee_{i=1}^k S^1$. Here $\pi_1(X) = G = F_k$, the free group on k generators. The universal cover $Y = \tilde{X}$ is a tree (as is the universal cover of any connected graph) and thus contractible, hence $X = K(G, 1)$ is aspherical. Furthermore, the cellular structure $e^0, e_1^1, \ldots, e_k^1$ shows that $\chi(X) = 1 - k$. By 6.1.5, and Definition 6.3.3, $\beta_0(X) = \beta_0(Y; G) = 0$. Hence, since $\beta_i(X) = 0$ trivially for $i > 1$, we have $\chi(X) = 1 - k = 0 - \beta_1(X)$, which implies that $\beta_1(X) = k - 1$, and $\beta_i(X) = 0$ otherwise.

Example 7.1.2. Let $\sum_g = X$ be an orientable surface of genus $g > 0$, with $G = \sigma_g := \pi_1(\sum_g)$ (e.g. $\sum_1 = S^1 \times S^1$, $\sigma_1 \approx \mathbb{Z} \times \mathbb{Z}$, and for $g > 1$ σ_g is an infinite non-abelian group). Here $\chi(\sum_g) = 2 - 2g$ is well known, and $Y = \tilde{X} \cong \mathbb{R}^2$. Again, $\sum_g = K(G, 1)$ is aspherical, and similar to the previous example $\beta_0(X) = 0$. In this case, since X also happens to be a 2-manifold, Poincaré duality (an ℓ_2 version, outlined in Appendix A to this chapter) shows that $\beta_2(X) = \beta_0(X) = 0$. Then $2 - 2g = \chi(X) = 0 - \beta_1(X) + 0$ implies that $\beta_1(X) = 2g - 2$, $\beta_i(X) = 0$ otherwise.

These two examples have an algebraic interpretation.

Definition 7.1.3. *Let G be a (necessarily finitely presented) group with a finite CW-model $K(G, 1)$, then*

$$\beta_i(G) := \beta_i(K(G, 1)).$$

From Remark 5.2.10 we can extend this definition as follows.

Definition 7.1.4. *Let G be a group with a CW-model $K(G,1)$ having finite n-skeleton for some $n \geq 2$ (called a group of type F_n), then*

$$\beta_i(G) := \beta_i(K(G,1)^{(n)}), \quad i < n.$$

In particular β_0, β_1 are defined for any finitely presented group (cf. §6.1).

7.2 *Deficiency of groups*

The deficiency of a finitely presented group G was defined in the Introduction 1.1.3. We now give the proof of the first easy result mentioned there.

Proposition 7.2.1. $\operatorname{def}(G) \leq b_1(G)$.

Proof. By definition $b_1(G) = \operatorname{rank}(H_1(K(G,1)))$, and

$$H_1(K(G,1)) = \pi_1(K(G,1))_{ab} = G_{ab},$$

since by definition $\pi_1(K(G,1)) = G$. Hence

$$b_1(G) = \operatorname{rank}(G_{ab}) = \dim_{\mathbb{R}}(G_{ab} \otimes \mathbb{R}) \geq g - r,$$

since $G_{ab} \otimes \mathbb{R}$ is given as a real vector space with g generators and r (linear homogeneous) relations. □

Our second result involves ordinary Betti numbers and refines the first proposition, then we turn to a few sample results with ℓ_2-Betti numbers.

Proposition 7.2.2. $\operatorname{def}(G) \leq b_1(G) - b_2(G)$.

Proof. For G given with g generators and r relations, we construct $K(G,1)$ as in §6.1, so $K(G,1)^{(2)}$ has 1 0-cell, g 1-cells, and r 2-cells. Then, using 6.3.9,

$$\chi(K(G,1)^{(2)}) = r - g + 1 = 1 - b_1(K(G,1)^{(2)}) + b_2(K(G,1)^{(2)}).$$

But $b_1(K(G,1)^{(2)}) = b_1(G)$ and $b_2(K(G,1)^{(2)}) \geq b_2(G)$, thus $r - g \geq b_2(G) - b_1(G)$, or $g - r \leq b_1(G) - b_2(G)$, as required. □

Theorem 7.2.3. $\operatorname{def}(G) \leq 1 + \beta_1(G)$.

Proof. Similar to 7.2.2, here we find that

$$\chi = r - g + 1 = \beta_0(G) - \beta_1(G) + \beta_2(K(G,1)^{(2)})$$

implies that

$$g - r = 1 + \beta_1(G) - \beta_0(G) - \beta_2(K(G,1)^{(2)}) \leq 1 + \beta_1(G). \qquad \square$$

We remark that in case $\operatorname{def}(G) = 1 + \beta_1(G)$, Hillman has shown [9] that the 2-dimensional CW-complex constructed from the g generators and r relations (as in §6.1) is in fact aspherical.

Corollary 7.2.4. *If* $\beta_1(G) = 0$ *then* $\operatorname{def}(G) \leq 1$.

Suppose $K(G, 1)^{(3)}$ is finite, i.e. G is of type F_3. The Morse inequalities 6.3.11 yield (with $k = 2$) $r - g + 1 \geq \beta_2(G) - \beta_1(G) + \beta_0(G)$, $g - r \leq 1 + \beta_1(G) - \beta_2(G)$. This proves

Theorem 7.2.5. *For any group G of type* F_3, $\operatorname{def}(G) \leq 1 + \beta_1(G) - \beta_2(G)$.

Note that any infinite fundamental group of a closed orientable irreducible 3-manifold M will be of type F_3, indeed a standard theorem of 3-manifold topology asserts that such an M is already aspherical.

Proposition 7.2.6. *Let G be a free group of rank k, then* $\operatorname{def}(G) = k$.

Proof. Since G has a presentation with k generators and 0 relations, $\operatorname{def}(G) \geq k - 0 = k$. But by 7.1.1 and 7.2.3, $\operatorname{def}(G) \leq 1 + \beta_1(G) = 1 + (k - 1) = k$. □

Similarly, using the standard presentation of σ_g with $2g$ generators $x_1, y_1, x_2, y_2, \ldots, x_g, y_g$, and the single relation $[x_1, y_1][x_2, y_2] \ldots [x_g, y_g] = e$, and using 7.1.2, 7.2.3 (7.2.2 may also be used), we have

Proposition 7.2.7. *For* $g > 0$, $\operatorname{def}(\sigma_g) = 2g - 1$.

Exercise. Consider the group $\mathbb{Z}*_m := \langle a, t | tat^{-1} = a^m \rangle$, $m \in \mathbb{Z}$. Show that $\mathbb{Z}*_m$ is solvable with deficiency 1. Note that the case $m = 1$ follows from Proposition 7.2.7 with $g = 1$, so one may assume $m \neq 1$.

We refer to [10], Ch. 2, Corollary 6, for the following theorem, which is related to the above exercise. It is a nice example of a purely algebraic theorem provable by ℓ_2 methods.

Theorem 7.2.8. *A finitely presented solvable group G has deficiency* 1 *if and only if* $G \approx \mathbb{Z}*_m$ *for some m.*

As a final application we mention a recent result of Kochloukova ([12], [13]), using ℓ_2 (and many other) techniques, that solves and even generalizes a thirty year old conjecture of E. R. Strasser concerning knot-like groups. A group G is called knot-like if it is finitely presented with a deficiency 1 presentation, and $G_{ab} \approx \mathbb{Z}$. The Strasser conjecture asserts that if G is knot-like and in addition G' is finitely generated, then G' is free.

Theorem 7.2.9. *Let G be a finitely presented group with a presentation of deficiency* 1, *and* G' *finitely generated. If* $f : G \to \mathbb{Z}$ *is surjective with finitely generated kernel* K, *then* K *is free.*

The Strasser conjecture is the special case where $G_{ab} = G/G' \approx \mathbb{Z}$ and f is the abelianization surjection. This case also has the following nice corollary.

Corollary 7.2.10. *Suppose in addition that* G' *is perfect, i.e.* $G' = G''$. *Then* G' *must be the trivial group and hence either* $G = \mathbb{Z}$ *or* $G = \mathbb{Z}^2$.

Proof. One now has G' both free and perfect, so it can only by the trivial group F_0 or $F_1 \approx \mathbb{Z}$. The result follows. □

7.3 *Amenable groups*

Let G be a discrete topological group and define

$$B(G) = \{f : G \xrightarrow{f} \mathbb{R},\ f \text{ bounded}\}.$$

Consider $B(G)$ as a left G-module by putting $(x \cdot f)(y) = f(yx)$ for all $x, y \in G$ and $f \in B(G)$.

Definition 7.3.1. *A mean on G is a linear map $M : B(G) \to \mathbb{R}$ such that for all $x \in G$ and $f \in B(G)$,*

(a) $M(1) = 1$ *(1 = the constant function 1),*
(b) $M(x \cdot f) = M(f)$,
(c) $f \geq 0 \Rightarrow M(f) \geq 0$.

Definition 7.3.2. *A group G is amenable if it admits a mean.*

This definition can be extended to any locally compact topological group, indeed to any topological group, we leave the (slightly technical) details for Appendix B of this chapter. A finite group G is amenable, indeed M is uniquely given by

$$M(f) = \frac{1}{|G|} \sum_{x \in G} f(x).$$

Similarly a compact group is amenable, replacing the sum in the preceding formula by the integral with respect to the Haar measure μ (normalized so that $\mu(G) = 1$.

The question of determining all infinite amenable groups is deep and has led to much interesting work. In particular every abelian and indeed every solvable group is amenable, a fact that is not obvious even for $G = \mathbb{Z}$. A relatively easy fact to establish is that a subgroup of a (discrete) amenable group is also amenable, this is left as an exercise for the reader. Elek [6] has shown that for a countable group G, $H_0^G(Y; \ell_2 G) \neq 0$ if and only if G is amenable, thus the result in Example 6.1.6 gives an ℓ_2-homology proof that $\mathbb{Z}$ is amenable.

Here, without proof, is a useful lemma of the theory (cf. [3], [5]), and some consequences.

Cheeger-Gromov lemma 7.3.3. Let Y be a connected free cocompact G-CW complex and G an infinite amenable group. Then

$$can^i : {}^{red}H^i(Y) \to H^i(Y; \mathbb{R}),$$

as defined in 5.2.12, is injective for all $i \geq 0$.

Corollary 7.3.4. *Given a finite connected CW-complex X with $G = \pi_1(X)$ amenable, then $\beta_1(X) = \beta_1(G) = 0$.*

Proof. Let $Y = \widetilde{X}$. Since Y is simply connected, $H_1(Y) = \pi_1(Y)_{ab} = 0$. It follows (universal coefficient theorem) that $H^1(Y; \mathbb{R}) = 0$, so by the Cheeger-Gromov Lemma ${}^{red}H^1(Y) = 0$. Thus $\beta_1(Y, G) = \beta_1(X) = 0$. Finally, since $X = (K(G, 1))^{(2)}$, $\beta_1(X) = \beta_1(G)$. □

In Example 7.1.1 we saw that $\beta_1(F_2) = 1 \neq 0$, which gives the next result.

Corollary 7.3.5. *The free group F_2, or any supergroup of it, is not amenable.*

Appendix A

Poincaré duality

The ℓ_2 version of Poincaré duality can be obtained as follows. Start with a closed manifold M, having dimension n, and let N be its universal cover. We suppose as usual that M, N have compatible cellular structures, i.e. the cells of N are p^{-1} of the cells of M. Also let $G = \pi_1(M)$ and for convenience write $R = \mathbb{Z}G$. Following [20], p. 215, there is an equivariant chain homotopy equivalence of (left) R-modules

$$K_i(N) \to hom_R(K_{n-i}(N), R).$$

Applying the functor $\ell_2 G \otimes_R \_\_$ to this gives a chain equivalence of the ℓ_2 chains and cochains $C_i(M) \to C^{n-i}(M)$ which in turn induces the ℓ_2 Poincaré duality.

Appendix B

Amenability

In §7.3 we worked with discrete groups G. More generally, let G be any Hausdorff topological group. Write $CB(G)$ for the space of all continuous bounded functions $f : G \to \mathbb{R}$, given the sup norm topology. A function $f : G \to \mathbb{R}$ is called left uniformly continuous if the map $G \to CB(G)$ given by $x \mapsto L_x f$ is continuous, and $LUC(G)$ denotes the space of such functions. We say that G is amenable if there exists a mean M on $LUC(G)$ satisfying properties (a), (b), and (c) of Definition 7.3.1. If G is locally compact, this is equivalent to $CB(G)$ admitting such a mean. Of course, in case G is also discrete, then $CB(G) = B(G)$ and this reduces to Definition 7.3.1. It is also worth noting, in the locally compact case, that the existence of a mean on $CB(G)$, i.e. amenability, is equivalent to the existence of a mean (as always, satisfying (a), (b), and (c)) on the larger space of essentially bounded functions that are measurable with respect to the Haar measure on G. This latter property is frequently taken as the definition of amenability in the literature.

References

[1] M. Atiyah, Elliptic operators, discrete groups and von Neumann algebras, *Astérique*, **32** (1976) 43–72.
[2] M. Burger and A. Valette, Idempotents in complex group rings: theorems of Zalesskii and Bass revisited, *J. of Lie Th.*, **8** (1998) 219–228.
[3] J. Cheeger and M. Gromov, L^2-cohomology and group cohomology, *Topology*, **25** (1986) 189–215.

[4] B. Eckmann, 4-manifolds, group invariants, and ℓ_2-Betti numbers, *L'Enseignment Mathématiques*, **43** (1997) 271–279.

[5] B. Eckmann, Introduction to ℓ_2-methods in topology, reduced ℓ_2-homology, harmonic chains, ℓ_2-Betti numbers, *Isr. J. Math.*, **117** (2000) 183–219.

[6] G. Elek, Amenability, ℓ_p-homologies and translation invariant functionals, *J. Austral. Math. Soc.*, **65** Part 1 (1998) 111–119.

[7] M. S. Farber, Novikov-Shubin invariants and Morse inequalities, *Geometric and Functional Analysis*, **6** (1996) 628–665.

[8] P. R. Halmos, Introduction to Hilbert Space and the Theory of Spectral Multiplicity, 2nd edition, Chelsea, N.Y. (1957).

[9] J. A. Hillman, L^2-homology and asphericity, *Isr. J. Math.*, **99** (1997) 271–283.

[10] J. A. Hillman, Four-manifolds, Geometries, and Knots, Geometry and Topology Monographs, 5, Geometry and Topology Pub., Coventry (2002).

[11] J. L. Kelley, General Topology, D. Van Nostrand Co., Inc., Princeton (1955).

[12] D. Kochloukova, On a conjecture of E. Rapaport Strasser about knot-like groups and its pro-p version, *J. Pure Appl. Alg.*, **204** (2006) 536–554.

[13] D. Kochloukova, Some Novikov rings that are von Neumann finite and knot-like groups, *Comment. Math. Helv*, **81** (2006) 931–943.

[14] P. Linnell, Division rings and group von Neumann algebras, *Forum Mathematicum*, **5** (1993) 561–576.

[15] W. Lück, Hilbert modules over finite von Neumann algebras and applications to L^2-invariants, *Math. Ann.*, **309** (1997) 247–285.

[16] W. Lück, L^2 Invariants: Theory and Applications to Geometry and K-Theory, Ergebnisse der Mathematik und ihrer Grenzgebiete, 3. Folge, Vol. 44, Springer-Verlag, N.Y., Berlin (2002).

[17] F. J Murray and J. von Neumann, On rings of operators, *Ann. of Math. (2)*, **37** (1936) 116–229.

[18] S. P. Novikov and M. A. Shubin, Morse inequalities and von Neumann invariants of non-simply connected manifolds, Uspekhi Mat. Nauk, **41** (1986) 222–223.

[19] F. Riesz and B. Sz.-Nagy, Functional Analysis, F. Ungar, N.Y. (1955).

[20] C. T. C. Wall, Poincaré complexes I, *Ann. Math.*, **86** (1967) 213–245.

[21] A. E. Zalesskii, On a problem of Kaplansky, *Soviet Math.*, **13** (1972) 449–452.

Made in United States
Orlando, FL
04 November 2026

90206046R00101